CORPORATE SOFTWARE
PROJECT MANAGEMENT

CORPORATE SOFTWARE PROJECT MANAGEMENT

GUY W. LECKY-THOMPSON

CHARLES RIVER MEDIA, INC.
Hingham, Massachusetts

Acquisitions Editor: James Walsh
Cover Design: The Printed Image

CHARLES RIVER MEDIA, INC.
10 Downer Avenue
Hingham, Massachusetts 02043
781-740-0400
781-740-8816 (FAX)
info@charlesriver.com
www.charlesriver.com

This book is printed on acid-free paper.

Guy W. Lecky-Thompson. *Corporate Software Project Management.*
ISBN: 1-58450-385-8

Library of Congress Cataloging-in-Publication Data
Lecky-Thompson, Guy W.
  Corporate software project management / Guy W. Lecky-Thompson.
      p. cm.
  Includes index.
  ISBN 1-58450-385-8 (pbk. with cd-rom : alk. paper)
  1. Computer software—Development—Management. I. Title.
  QA76.76.D47L42 2005
  005.1'068—dc22
                              2004025343

Printed in the United States of America
05 7 6 5 4 3 2 First Edition

CHARLES RIVER MEDIA titles are available for site license or bulk purchase by institutions, user groups, corporations, etc. For additional information, please contact the Special Sales Department at 781-740-0400.

Requests for replacement of a defective CD-ROM must be accompanied by the original disc, your mailing address, telephone number, date of purchase, and purchase price. Please state the nature of the problem, and send the information to CHARLES RIVER MEDIA, INC., 10 Downer Avenue, Hingham, Massachusetts 02043. CRM's sole obligation to the purchaser is to replace the disc, based on defective materials or faulty workmanship, but not on the operation or functionality of the product.

This book is for my son, William,
who slept and cried his way through most of its writing.

Contents

Preface

It has long been accepted in the software industry that projects will be late, over budget, and lacking in agreed features due to technical or time limitations. In part, this has been due to the neglect of the people involved in the design and implementation of large-scale software development projects. This problem is compounded by what has become known as the "intangibility" of software. It cannot be touched, felt, or viewed in a physical way—it exists only as an abstraction of what it represents.

All these problems taken together conspire to make the project appear more complex to manage than perhaps it should be. By using rigid communication lines, and well-defined relationships between different parties, much of the complexity that stems from the need to integrate different people's specialties can be removed.

This leaves management to worry only about what must be done at a technical level, and improves relationships between different departments. On top of this, there is also an increase in client-corporation communications, which in turn, reduces the possibilities of making errors in judgment that can be costly to track down and fix later in the project.

Besides the rigidity of the approach taken in the book toward managing large-scale development teams, there is also the built-in flexibility to define standards at a corporate level. Guidelines are given as to the type of standards that need to be created, and what they should cover, but the actual implementation details are left largely to the corporation's own staff members.

Aside from the management aspects, the technical side of software engineering is also dealt with. This part of the book illustrates the decisions that need to be made when planning the implementation phase of the project. In addition, there are also examples of ways in which the software can be designed so that the reuse of code is maximized for future projects.

Throughout the book, the focus is on three main points: management, quality, and client relations. The key to success in the corporate software engineering field is in effective management of the tasks that need to be performed, ensuring that they are correctly performed, and that they accurately reflect the needs of the client.

This book presents methodologies and paradigms that can be applied in the real world to ensure that these three main aspects are catered for. However, no book on this topic would be complete without a discussion, however brief, on the technical issues behind implementing the designs agreed on with the client.

In recent years, models and paradigms for software development have stabilized, and in some cases stagnated. Computer language development has likewise reduced to the point that enhancements are being made infrequently, and the language cores are not changing at all.

This latency in the market means that for the first time, concrete guidelines can be set out that will not change as innovations become available to the general public, in the way they once did. As more and more companies become involved with the development and deployment of large-scale software projects, it has become necessary to define ways in which control methods can be applied to ensure that a quality product results.

In *Part I: The Product Development Mix,* we deal with the relationships between the various teams that contribute to the finished product. We concentrate on offering new insights into the way process control and communication in the area of corporate software engineering can be achieved by attempting to break the traditional top-down team structure into a collection of intercommunicating units.

One of the key principles is in using a central point of contact, coupled with clearly defined processes and responsibilities, and well-documented process control defined by rigid guidelines. Much of the actual implementation decisions are left with the organization itself; this book gives an indication of best practice but is careful not to impose rigid solutions. That is up to the way that the implementation is carried out.

What is also important is that the organization realizes that the techniques must be used in a way that is conducive to good results, taking into account the relative size of the organization. It makes no sense for the management of the processes required to be allowed to grow into a resource that is the greatest part of the company—if this means that some staff need to perform two functions, so be it.

The central theme of this book is described in *Part II: Principles of Corporate Software Engineering,* which uses material developed in Part I to establish the interactions required to specify and implement a high-quality end product. Here again, we concentrate more on the processes that need to be in place, and a template for their implementation, than we do on ground-level technical issues. It is assumed, for example, that the programming team knows how best to implement a design, and is equipped with the tools to do so. Any information pertaining to design and programming is given at a high level.

With this in mind, it is not necessary for the readers to be fully acquainted with the intricacies of software development, since enough introduction is given for them to be able to follow, manage, and set up a well-functioning development center.

Part of the innovation in Part II revolves around code reuse and taking advantage of new paradigms such as Open Source software and advanced scripting languages for integrating existing components. Indeed, in the same way that communication and standards are the key to Part I, reuse and documentation are key to understanding the driving force behind Part II.

Finally, in *Part III: Principles of Software Quality Control*, we discuss ways in which the quality of the end product can be ensured and the steps to take if the client becomes dissatisfied in some way with the progress being made. Quality Assurance and Control usually gets short shrift in software engineering books, since the assumption is that adequate testing is enough to be able to deliver a high-quality result.

Where corporate software engineering is different is in realizing that software creation today is not so much production, as delivery of a service. As such, it is only fitting that some service quality assurance paradigms be adopted for the control and auditing of the quality level of an organization involved in software engineering as their primary business. In addition, in Appendix B, "About the CD-ROM," the various resources are listed:

- Tools and Resources for project and quality management
- Templates for use in real-world situations

Part

I

The Product Development Mix

This part of the book centers on the teams that will be involved in taking an idea and turning it into a product. The initial idea may come from a response to a Request for Proposals (RfP), or it may be an internal idea that services an existing market need.

In either case, certain procedures can be put in place, and an infrastructure can be installed that helps to perform the tasks needed to create a well-conceived, high-quality product based on the initial input.

Much of the details are common to both on-spec projects and those created with a commercial venture in mind. It is important from the outset to realize that there are two types of client—internal and external—and that both are equally important in the product development life cycle.

The various theories and solutions that are put forward here exist in other industries, but the aim of this part of the book is to gather and reuse techniques in a way that is appropriate for use in a high-tech industry such as software development.

In Chapter 1, "The Liaison Center," the reader will learn about the way in which different teams can communicate with each other through a central point of contact, and how this can be used to achieve higher quality results in a more efficient manner.

Chapter 2, "Standards and Guidelines," lists various ways in which the organization can be sure that every team member performing a similar task is doing so with respect to a set of organization-wide standards. These are important in order to perform the various tasks in the most efficient manner possible, including effective communication.

In Chapter 3, "Specifications," we discuss how to define the product in terms that can be understood by technical staff and the end user, and are yet unambiguous enough that no misunderstandings on either side occur. It also provides a basis for ensuring that, using a combination of communication and standards, the process of developing the specifications yields a high-quality return early in the process, which will enhance the final delivery.

Chapter 4, "Product Development," pulls together the previous three chapters into a model for software development that is scalable in both directions, and adheres to existing guidelines for producing software.

Finally, Chapter 5, "Testing," defines how, after development has started, different testing methodologies can be applied to ensure that the result is worthy of the investment made in it.

1 The Liaison Center

In This Chapter

- Introduction
- The Role of the Liaison Center
- Key Tasks
- Personnel
- Implementing the Liaison Center
- Supporting Media
- Summary

INTRODUCTION

Software projects often comprise many and diverse teams, all with their own schedules and goals. The key to effectively managing the relationships between all the teams involved in software development on a large scale is establishing a central point of contact, the structure of which is outlined in this chapter.

Before we begin with our analysis of the ideal structure for the Liaison Center, we should first take a moment to look at what teams might be involved during the Software Development Life Cycle (SDLC). While this may vary from company to company, it is probable that the reader will find a comparable entity for each of the teams that we list here—albeit under a different name. In trying to list teams in this way, it is also inevitable that we will miss some, misname others, and arrive at a model that may not fit exactly into the organizational paradigm employed by a particular software engineering company, but this is also due to the intangible nature of software. In other production environments, such as the automotive or electronics industries, where a tangible product emerges from the production line, it is often easier to round up the teams responsible for the manufacture of the item in question.

Quite often, it is almost a "pick 'n mix" approach that is used to build a Software Engineering team, taking only those people required to produce the desired results, without actually creating a formal framework to drop the staff into. This is best avoided, and a rigid structure put in place by areas of responsibility with clearly defined boundaries. In this way, a more formal team structure can be built up with reporting lines that can be documented and followed.

While there may be a temptation to read this chapter as if the paradigms and processes that we discuss apply to a large organization, it is assumed that the various pieces that make up the Liaison Center are assembled in a manner that fits the resource availability of the target organization.

In other words, some of the functions will need to be appended to the job descriptions of other staff members, and in many cases the roles will be fulfilled in a way that may be secondary to their roles as secretarial, technical, or management staff.

It is the introduction of the controlling processes that are managed by the Liaison Center that is important, not the staff operating it. Indeed, it is more of a virtual department than an actual office in the building. We need it because communication on software engineering is important, and because in a small organization, staffed by technical personnel, there is a danger that it will become neglected.

In a large organization, other pressures are at work, which means that the Liaison Center will be more of an actual office, and less of a virtual structure. Nonetheless, the principles that apply to a small organization will also apply to the larger ones. It is a concept, and as such, is scalable.

THE ROLE OF THE LIAISON CENTER

One of the most important concepts behind the Liaison Center is that it provides a communication hub between all the different parties that will be involved throughout the development cycle of the product. This is not restricted to any given project, since the Liaison Center may be involved in many different projects at any given time, and will be responsible for maintaining good communication across departments and areas of responsibility.

Since the Liaison Center staff are responsible for much of the inter-project communication in the target organization, they also need to be appraised of all relevant project information. Thus, it also makes sense for the Liaison Center to manage the information flow and storage of that information. Since the Center holds all of the project information, the role will also include a certain amount of management responsibility.

The first level of responsibility is in scheduling the project staff, both across projects and within each project team, in accordance with both the wishes of the client and the management of the organization itself. Above this, there is a certain amount of resource management with which the Center will become involved, a logical progression from performing the scheduling aspects of project management.

At the highest level of responsibility for the effective central management of all projects is the contract and client management. Contracts should pass through the Center, since they will know if the terms can be met in a timely fashion, and what cross-pollination between projects (if any) can be achieved.

In a large organization, where many projects are being undertaken, it is clear that the role of the Liaison Center will require at least one, and probably more, full-time member working to ensure that the communication between projects remains intact. We assume, however, that smaller organizations will be working on fewer projects, and therefore that the role outlined previously can be split across multiple staff members.

Internal and External Clients

Sometimes, project teams will be servicing other project teams, all working toward a common goal. These should always be treated as *internal clients*; that is, they must be treated with the same level of respect as other, *external clients*. In this way, we can be sure that the work that is performed is of the same quality across the board—be it high or low. In large organizations, running multiple projects, some teams might not even know if the client is internal or external.

Programmers reading this probably think that *they* would know if they were working on a piece of code for an internal project simply by the nature of the task that they have been asked to perform. In a similar fashion, if it is just a piece of code to be bolted onto something else, then the task will be of a different nature than writing a full application (or even application "glue" to hold multiple components), thus it must be for an internal client.

This way of thinking can be eradicated by using the Liaison Center concept, and effectively modular code. Each piece of software should be viewed as a collection of components, all glued together by a piece of code that gives the external appearance and interface to the application. If the Center is used to its maximum potential, along with the other aspects of the Product Development Mix that we detail in this part of the book, then it should be almost impossible for a given team to tell if they are working for an internal or external client, removing the temptation to treat them differently.

Of course, the smaller the organization, the less this will apply, and it is likely that every client will be an external client. However, each programmer is creating something (or reusing something) that is designed to be integrated with something that has been created by somebody else. This means that, without explicitly realizing it, the Internal Client role has been fulfilled, and using the Liaison Center as a concept rather than a physical body can help deal with this aspect of software creation.

KEY TASKS

Having detailed the role of the Liaison Center, and how it fits in with the strategic clients and the vague form that it might take, now we should try to isolate what falls into the scope of responsibility linked to the Liaison Center, and how these tasks might be accomplished.

One important facet behind the Liaison Center is that it must be flexible enough to take on additional tasks where necessary, and yet relinquish some where they might not be needed, or indeed relevant to the role of the Liaison Center within the target organization. In other words, there will be times when particular tasks otherwise assigned to the Liaison Center might be better given to the project team proper or simply not performed at all.

This will also be the case in small and medium organizations using the lowest scale Liaison Center: as a concept, and not an actual department. In such cases, the key tasks that we outline need to be placed under the responsibilities of staff members in a best-fit approach.

For example, communication between clients and technical teams might be best interfaced by a member of the central secretariat, or an accounts manager. Leaving such a task to a highly technical person may result in communication difficulties, since technical staff do not generally communicate very well with nontechnical staff in a working environment.

Client Communication

The Liaison Center provides the first point of contact between the client (be they external or internal) and the project teams involved in serving the wishes of the client. In a traditionally structured organization, this will typically fall to the Project Manager. By transferring some of the communications away from this power base, it is possible to add an additional reporting level between the project team and the client.

This has two direct consequences. First, the client only deals with a single point of contact, which is a bonus, because over long projects, or those with problems, the project team, including the manager, may change.

The second consequence is that a new level of complexity is introduced. Normally, this is to be avoided within software engineering projects, but we will spend a large proportion of this chapter seeing that by adding complexity, and by default, reporting levels, we can retain a hold on the project by controlling each aspect separately. This is in stark contrast to the way in which some smaller software houses operate, where the project manager is burdened with all the responsibility of management and clients alike—often resulting in conflicts of interest.

Scheduling

One of the principal pieces of information that clients often want to be informed about is the progress of the project. Nothing is worse than the client being kept in the dark regarding how much, or indeed how little, work has been performed, and how much remains to be done.

When there are several teams involved—often with one controlling Project Manager, several Team Managers, and a collection of regular programmers, documenters, and even artists in some cases—scheduling the visible aspects of the project toward the client can sometimes become obscured.

In severe cases, even the project teams themselves might find it difficult to know the exact state of the project. Clearly, taking away this aspect of the engineering task will ease both their jobs, and improve the efficiency of the entire endeavor.

Therefore, the Liaison Center can also be used to translate the wishes of the client into scheduled tasks with clearly defined milestones and target dates. Close monitoring of these targets can then be converted into information that can be relayed to the client, which will ease their minds, or at least allow them to reflect upon decisions that might need to be made regarding the ongoing project.

Such decisions might result in reducing the overall complexity by altering the nature of the deliverables according to the amount of progress being made, or indeed allowing bonuses to be paid for advance completion of various milestones, or the entire project.

Resource Management

Coupled with scheduling the projects that are being serviced, there is the allocation of resources needed to achieve the goals set by the scheduling requirements. Smaller organizations will find that if they concentrate solely on one large project, they may find the infrequency of payment that this entails forces them to overspend and increase corporate debt levels.

To counteract this, they may find that taking on smaller projects with some synergy toward the larger, more lucrative (long-term) projects becomes the only way to increase cash flow to the point that the targets can be reached. This includes reducing staff idle time across projects and project teams, so that one project team might complete one task and begin another, while waiting for other teams to reach their own milestones.

Knowing which staff are available for which tasks is another responsibility that falls to the Project Manager, who suddenly finds that he is in demand as the number of smaller projects increases. This will place unnecessary strain on the Project Managers, and in turn reduce their efficiency. Thus, the Liaison Center can help reduce their workload by simply requesting resource information from them, and shuffling work around to meet all the milestones in place under the various contracts being serviced by the organization as a whole.

Team Coordination

Each project team will have its own Project Manager. Larger teams will be split into subteams, with their own Team Leaders, reporting to the Project Manager. Following existing practices, each project team will typically coordinate via the Project Manager to check that their statuses are conducive to the timely completion of the project on which they are working.

This approach is perfectly satisfactory when there are a few project teams working toward a common goal, but in the days since the hi-tech bubble burst, it is simply not profitable enough to sustain growth in a diminishing market. As noted previously, larger projects must often be subsidized by smaller ones, in an effort to match incoming funds with outgoing costs without increasing debt. The harsh reality is that investment in software firms is becoming difficult to come by as the number of industry failures rises.

Following the Liaison Center approach, it also becomes apparent that the coordination between teams, across projects, with reference to available resources and schedules in place, should be performed centrally. Those making the decisions become those with the best information, and those who are most appropriately informed about the status of all projects being serviced by the target organization.

Project Database Maintenance

Besides the actual deliverables that the project teams should be required to generate as a matter of course, a fair amount of information about the projects will be generated during their life cycle. There will be costing information, including staffing and equipment, all the information relating to team efficiency, and personal development of staff members during the life of the project, much of which is passed back to the Liaison Center from the various project and team managers.

This information needs to be collected, referenced, and stored in an appropriate manner. If ever there comes a time when other departments within the organization need some guidance when preparing quotes, or deciding to move into a new market, the Liaison Center should be the first place to which they turn.

Even the human resources department might require some data on staff performance, to help in rewarding employees for a productive year, if the company is doing well, and perhaps administering suitable penalties toward those unproductive staff members should the need arise.

The database should be able to allow ad-hoc queries of this nature, as well as some form of routine reporting that will be used as a guide to the general health of the software engineering department. It will need to store various pieces of information, both quantitative and references to qualitative data, probably indexed in a manner that allows easy retrieval.

Naturally, the role of the Project Database is one that in smaller organizations can be partially automated. In other words, the smaller the sets of data that are introduced (probably by various people), the easier it is to manage without human intervention. Where large numbers of projects are involved, it becomes necessary to employ a member of staff to clean up the collection of data and ensure its correctness.

Knowledge Management

Besides the information that is project related, and entirely a result of passing empirical data between the project teams and the Liaison Center, there will also be a vast amount of knowledge gained by the employees as they perform their daily tasks. In cases where the organization is acquiring large amounts of specialist knowledge relating to a niche field, the question often arises as to what the effect on the project would be, should an employee become suddenly unavailable.

Ideally, the employee should be replaced within a short timeframe, but the reality is that this will probably prove impossible. Instead, there is a requirement to store reference points to the knowledge that is being accumulated, such that retraining of an employee with the correct background becomes a much less arduous and time-consuming task.

Contract Management

Finally, it is necessary to ensure that the contracts that have been signed are being honored. Someone also needs to know when contracts are due for renewal (maintenance contracts, for example), or closing contracts that are no longer required.

Then there is the signature management itself—a standard contract might need four separate signatures, and if work has begun before all four parties have signed, then it may not be legally binding.

It may seem to be a trivial task, but again, if the Liaison Center paradigm is to be followed, it makes sense to ensure that those in possession of all the facts regarding progress and scheduling are those who actually watch over the contracts that have been signed, and enforce them where necessary.

As a bonus, the client still only has one point of contact, even at this level. Such transparency of communication enables the client to develop a rapport with the target organization that is not possible if they are constantly trying to communicate with several heads of the corporate monster.

PERSONNEL

The role and tasks performed by the Liaison Center must be driven by its personnel. Choosing staff to manage the Liaison Center, whether it is in a small organization employing the conceptual (virtual) approach or a real office within the organization, will be constrained by the departmental and management model of the target organization. This is because a one-project outfit will probably incorporate much of the work of the Liaison Center within other project groups and subgroups.

It may not be profitable or desirable for the target organization to separately staff and run an effective Liaison Center, preferring instead to allow the projects to liaise together without a formally staffed structure in place. Of course, various secretarial duties must still be performed, including maintaining a professional and effective audit trail and full set of documentation.

Whether the Center itself is a separately managed business unit or a shared responsibility, it is important to ensure that the staff chosen to perform the duties outlined in this chapter do so effectively. Software engineering processes in general, and in the corporate environment in particular, are akin to a machine comprised of parts having a delicate relationship that is based on reliance on each other. If one breaks, or is missing, the entire machine functions at a less effective level—often leading to expensive mistakes.

While some of the responsibilities will be shared, three main areas need to be serviced, which reflect the role and the tasks that the Liaison Center needs to perform.

Communication

One of the most difficult tasks is communicating toward both the client and the employees in a manner that achieves the best results. One has to be firm toward the employees, but give credit where it is due, and at the same time pacify, when times are bad, or appear meek toward the client when times are good.

Choosing an effective communicator is as difficult as the task itself. Not many people have the mix of a technical background with effective communication skills. This may be why much software documentation is very hard to read for most non-technical users (keeping a professional tone; less humor).

The result of this paradox is that it may just be easier to pick a good communicator, and train him in the technical side of the job while he is performing it, rather than try to make a good communicator out of a highly technical specialist. If a budding human resources manager is lucky enough to find such a gem, she is advised to hold onto that person—he is a valuable commodity.

Toward the employees of the organization, it is often tempting to underplay the importance of diplomatic communication skills. The usual attitude is that they are there to perform a task, and they should get on with it, with a little praise if they do well, or often an outright verbal assault should they make a mistake.

Admittedly, these are extremes, but almost never the correct approach to take. Never mind the oft-repeated adage of "a little praise goes a long way"; most technical employees will require gentle handling. This not being a textbook about human resources, or employee psychology, many different tacks can be taken when dealing with staff from a management point of view, but the Liaison Center communications staff will not necessarily be well versed in advanced management skills either.

In short, then, effective communicators should be obvious in interview, and even role-playing, or in their regular job. If they are not, then they may not be the ideal choice for the position.

Documentation

Much of the Liaison Center's responsibilities revolve around producing, reading, storing, analyzing, and generally managing various pieces of documentation. Some will, as we shall see, be standard, company issue documents that have a structure that has been well defined in advance. Others will be ad-hoc documents that need to be read and understood, or converted to adhere to those same company standards.

Clearly, one of the members of the Liaison Center, whether it is his primary function or in addition to his usual tasks, needs to have an affinity for the written word. A merging of the tasks allocated to him as a member of the Liaison Center and his regular job would be ideal, in cases where the Liaison Center is a shared responsibility.

Of course, if setting up a Liaison Center becomes a company objective, it is slightly easier to recruit someone who enjoys writing, reading, and organizing documents. With the aid of the standard templates that we will discuss later, and some software, the person responsible for documentation needs to be able to store, retrieve, and report on all manner of documents supplied by both internal and external sources.

Management

Some fairly high-level issues can only be handled by "management." However, this does not necessarily mean that the persons serving the management role in the Liaison Center need to actually hold the same status as the manager of Human Resources, or MIT, for example. However, it will help if there is a member of staff with the respect and experience accorded to a senior member of the organization.

Within the Center, the management is responsible for ensuring that their staff perform in an effective and productive manner, and that there is effective communication within the Center. Information is the key to the success of the Liaison Center; therefore, the flow of information needs to be carefully monitored, and any shortcomings dealt with in a timely manner.

The personnel selected to manage the Center need to service the relevant tasks with a clear management perspective—in terms of scheduling (and time management) and resource and contract management skills.

Thus, they need to have relevant experience, a clear sense of judgment, and above all, the ability to communicate effectively with staff, management, and clients.

IMPLEMENTING THE LIAISON CENTER

From the three previous personnel descriptions, it is clear that much of the work that the Liaison Center does is actually nontechnical, although all staff should have a basic understanding of the processes involved in creating software. Even if the Center becomes a shared responsibility, the skeleton service should include three full-time members of staff as a minimum—a secretary, librarian, and manager—although these can be shared with related tasks in nontechnical areas.

Depending on the protocols in place in the target organization, these three should be placed in an open-plan office, in an attempt to increase verbal communication. There is nothing worse than isolating staff involved in the tasks outlined in this chapter in little cubicles. This is only, however, the opinion of the author, based on his own observations.

Staff can be added to the Liaison Center in an ad-hoc basis as the needs arise. In certain circumstances, the secretarial duties and documentation management tasks can be merged, leaving a staff of two. This approach should only be used when the number of projects is small enough that the workload would not be sufficient to occupy two separate members of staff.

IT Infrastructure

Usually, the target organization will have certain policies that dictate the distribution of IT equipment, from workstations to servers, networking equipment, and so forth. However, due to the quantity and nature of the information that will be accumulated, stored, and referenced, it is advisable to allocate a separate server for the Liaison Center, with sufficient storage capacity, backup procedure, and robustness.

If electronic mail is used for communication within the organization, mailboxes should be created for project management and documentation. These will be used as a depositing point for information pertinent to the progress of projects, and associated documents. On the server, a folder needs to be created for each project that is being serviced by the organization.

In the course of this book, we will be adding information stores and subfolders to this initial structure, assuming that the bare bones of the IT Infrastructure are in place.

Selling the Liaison Center

Currently, times are hard in the software development industry. In such times, it can become increasingly difficult to sell something that has every appearance of being a cost center as opposed to part of the revenue machine that keeps the organization afloat.

However, it is a cost saver; without it, the software development process can be in danger of losing its way, and individual software engineers, designers, and programmers may not be working within an infrastructure that makes the best use of their time.

Part of the guiding philosophy of software development that we will be covering later in this book relates to practices promoting code reuse as a means to reduce time to market and increase overall product quality.

Managing code reuse and the standards that need to be put in place to ensure that it is conducted in a manner that leads to satisfactory results requires that there are members of staff responsible for both the artifacts themselves (being both documentation and code) and the supporting infrastructure.

If the production of software is the core business of the target organization, then creating a team of two people (as suggested previously) as a way to reduce the substantial costs associated with quality failures in the software industry begins to make sense. It should be presented as facilitating an improved software development paradigm.

The other side is the client-facing emphasis that the Liaison Center promotes as a key working principle. If the organization is involved in any way as an outsourcing partner, then it will benefit from the added processes that are in place to ensure effective communication between the project team and the client.

Added to this is the possibility to realize synergies between or across projects, which leads to even greater cost savings. Arguably, it would not be possible to achieve these synergies without the guiding hand of the Liaison Center.

SUPPORTING MEDIA

All of this relies on being able to effectively pass and store information between the involved parties. To facilitate this, it is necessary to devise a system of standard templates for documents that can be used to transfer knowledge from team members to the Center and on to clients where necessary.

Industry Standards

There are a number of standards used in the technical writing industry, ranging from simple word processor template sets that can be reused in corporate documentation guidelines, to complete document engineering standards that require substantial training to use effectively.

They are all based on one simple principle: the ability to effectively identify the problem domain (that which we want to document) and then break it down into pieces that can be logically grouped in order to create a structured final document. This is much like the mechanisms that are needed to perform software engineering, and so it is logical to assume that good software engineers can also make good document engineers and technical writers.

Unfortunately, where this connection falls down is in the realization of two basic differences between technical writers and software engineers or programmers. The first is that programmers are generally not very good at conveying complex ideas to a third party as a collection of simple ones. They assume that the reader has the same level of technical competence as they do, which is not necessarily going to be the case.

The second difference is that, unlike technical writers, programmers and software engineers do not enjoy writing documentation, since they would much rather be writing programs, designing complex systems, or debugging—anything other than writing documentation.

It is, however, easy to use software engineers in an informal workshop situation to help the technical writers to break down the problem domain and organize the general structure of the document that is to describe the system that needs to be put into place. It is basically the same set of skills, and much the same work needs to be done to engineer the system and to document it.

Even if the investment in a specific industry standard is deemed out of reach for the target organization, there are still basic principles that can be adhered to when creating documentation that revolve around psychological research, such as the fact that the human mind can deal with about five pieces of information in short-term memory at a time.

Thus, the document should be written in such a way that in order to understand a given key concept, it should not require the readers to hold more than five other concepts in their minds at once. These supporting concepts may be key, or they may be throwaway pieces of information that enable understanding but can be discarded almost immediately once the concept that they support is understood.

If one follows this theory to the letter, any system should be broken down such that it is comprised of five or less key concepts, each of which can be explained until comprehension by five or less supporting concepts, which, in turn, can be supported by five or less subsidiary concepts, and so on, through the system.

Another accepted principle of document construction is that no one idea should require more than a single page to express. Moreover, each page should only consist of a certain amount of text—techniques for cheating, such as reducing the font size to an almost unreadable 8 points, narrowing the margins, and removing header and footer information are all strictly forbidden.

The idea is that the coverage should not exceed one-third to one-half of the paper width, or three quarters of the available page height, after the printing margins are taken into account. The rest of the space is divided into four areas: the header and footer (which contain vital document information), a left-hand margin designed to contain labels for each concept and subconcept, and a right-hand margin that is wide enough for the reviewer to write notes in.

If these principles are borne in mind, the resulting document should be well thought out, easy to read and interpret, and communicate the information in a way that makes understanding less of a chore and more of a pleasant reading exercise.

Layout and Document Structure

A structured document should consist of the following sections:

- Document Information
- References
- Table of Contents
- Glossary of Terms

- Main Document
- Appendixes where appropriate

The document information section is there to inform the reader as to the audience, subject material, layout, and style of the document, and as such needs to include headings such as:

- Change History
- Overview
- Section Summaries

The Change History subsection needs to contain details of each revision of the document, along with the reason for the changes, date, and author. The overview simply addresses the intended audience with a management summary of the information contained within. Each section summary is then a single line designed to convey, in one sentence, the exact nature of the information it contains.

Following on from the Document Information section is a list of other documents that are referenced within this particular work. Each entry should follow the accepted standards for citing sources that the organization has chosen to adopt.

The Table of Contents, following the References, will likely be the result of an automatic feature of the word processor used to edit the document, as will the Glossary of Terms, although the latter will need to be annotated by the author of the document. Any acronyms used in the document should also be referenced in the Glossary.

Following all the introductory sections is the main document, which can be laid out in a variety of different ways, although it is best to retain a Chapter, Section, Subsection approach, with a depth of no more than three levels. This is a flexible rule of thumb that the author has found to work well when constructing technical documents with the help of software engineers and information management staff.

Finally, there are the appendixes, which need only be present if there is information such as forms, diagrams, or software user manuals that are presented as supporting documents for the information presented in the main document section. It is usually easy to decide whether a particular piece of supporting documentation should be put in an appendix. As a rule of thumb, information that supports a decision presented in the document but is not required in order to understand the effect of that decision should be placed in an appendix.

Information Management and Document Information Systems

The reason for investing time and effort in creating a collection of documents for use in the corporate environment—whether they be specifications for software systems, user guides, service descriptions, legal references such as contracts, or simply descriptions of modules or object code that exists for use in future projects—is so they can be referred to when a decision needs to be made.

This is useless if the documents cannot be found. Eventually, such a complex mountain of documentation will exist that, should a single document need to be found, a good indexing system had better be in place; otherwise, it will be like looking for a needle in a haystack.

Therefore, it is vital that the indexing, referencing, and search mechanisms are put into place before the document collection is begun, since trying to retrofit a good system after the fact will be expensive and the results will be less than satisfactory.

Should the readers find themselves in the situation where there is a vast mountain of documentation that is unsorted, badly indexed with no possibility to search by keyword or perform a full-text search, it will be much more efficient to begin a new document archive, and slowly bleed the existing documents into it, after they have been reformatted.

Each document needs to contain a set of keywords that identify the subject matter in an abstract fashion. If the word processing package does not support document properties that enable this do be done in a way such that they do not appear in the text, then a separate text document containing the keywords should be created with the same base name as the document to which it refers.

In this way, a person who needs to locate information on a given topic can perform a two-tier search: first by keyword, resulting in a set of likely candidates, and then a full text search to unearth the exact information he might require. This has been found to work with greater efficiency than a full text keyword search by itself or a simple keyword indexing system.

With the advent of the Internet and World Wide Web, much research into document organization and search mechanisms has been performed, in an attempt to create the best search engine, capable of finding the most relevant Web pages to display to the user. One of the by-products of this research is the glut of various search systems that are available for document organization.

Good though these are, in the end, the title of the document is one of the best ways for someone browsing the document archive to find the information she is looking for. Hence, organizations should be wary of using strictly numerical indexing systems. In the past, filenaming conventions on operating systems have led to restrictions (such as the DOS 8.3 format), which meant that giving filenames to documents that conveyed meaning was impossible.

With long filename support existing in almost every operating system in the field today, there is no excuse for using strange numerical indexing when we can all use plain-text names that will help the user determine whether the document is likely to be of interest. In creating the document archive, this should be taken advantage of fully. Be aware, however, that if a backup to CD-ROM of the archive is to take place, certain restrictions in the CD-ROM format mean that some file-names will probably be truncated.

SUMMARY

This chapter laid the foundation for the rest of Part I of the book, in the sense that it outlined the container into which the information generated by implementing the rest of the Product Development mix, and their associated teams, will be placed. In addition, it provided the coordinating facility that underpins the key processes that support creating software in a scalable fashion.

The preceding statement is important, since most companies involved in creating software start out small. Even Microsoft began as a two-person operation. At some point, it will become necessary to implement controls that will increase the effectiveness of the operation. This includes increasing quality, reducing costs, and operating profitably enough to promote expansion.

Without a good operating structure, one of three possible scenarios could ensue. First, the company could fail to operate effectively, and thus not survive. Second, the company could grow to a size that is not manageable within the confines that bind it, and implementing a suitable structure once the organization has expanded might prove expensive.

The final scenario is one in which the quality of the software created by the company does not meet client expectations, or in which the efficiency of the production process is impaired such that, although the company survives, it cannot expand in an efficient manner. This may or may not actually matter; however, most heads of companies prefer growth to stagnation.

The Liaison Center may seem like a simple concept, even redundant, but it does provide an alternative framework that can remain an effective anonymous control and coordination point in the organization, leaving the technical staff free to perform their tasks unfettered. This is also the case where it is they who are providing some service to the Liaison Center; the processes and structures involved will mean that they will spend less time worrying about issues dealt with by the Liaison Center than if it was not there and they had no template for performing those tasks.

2 Standards and Guidelines

In This Chapter

- Introduction
- Defining Standards
- Project Documentation
- Coding Standards
- Data Collection Standards
- Reporting Templates
- Summary

INTRODUCTION

Without corporate documentation standards, communication and information storage in an effective, efficient, and profitable manner will prove difficult, if not impossible. Standardizing the kinds of information to be stored will mean that data entry can be automated (at best), or rendered much more efficient (at worst).

In addition to standards within which information can be gathered and stored, there should also be guidelines that detail the processes and procedures governing the collection of the data, along with ways of tracking what has been collated, and reporting to other parties.

It is also important that the standards be used to improve readability and interpretation. No matter who has written the document, the client should be able to recognize the culture of the target organization through the documentation that has been produced. They should not, for example, need to try to maintain a list of ambiguities by document author—the style guide used to write the documents should remove any possible ambiguities if followed by all staff.

DEFINING STANDARDS

Within the documentation, there will be specific ways in which the information should be represented; and the way in which we define this presentation is detailed in a single document that indicates the standards that must be adhered to when reporting within the target organization. There are many reasons for defining a set of reporting standards, but key reasons include ensuring that there is no ambiguity on the side of the reader and the author of the document.

Since miscommunication is one of the main causes of project failure in the early stages, it is important to remove any ambiguities in the documentation. However trivial some of the standards that need to be defined may seem at first sight, over-looking some of the details could lead to problems with far-reaching consequences.

The following are simply suggestions, and the target organization may want to adopt different conventions than those listed here. This is not a problem, as long as it is clear that there is one way to represent information in a given situation, and that the guidelines are published to all staff responsible for writing documents, memoranda, or even internal messages. Currently, this will probably include all staff.

Date Standards

Since documents are often referenced by date, it is important that document titles, section titles, and references to decisions or meetings all follow the same format. In general, there are two frequently used styles: the European and U.S. variants.

In Europe, the accepted short form is Day-Month-Year, as in 31-12-2004, with titles and ad-hoc references being of the form 12 December 2004. This is different from the U.S. convention, which is usually Month-Day-Year (12-31-2004), and December 12, 2004 for the longer form.

Since the year 2000, many organizations have abandoned the two-digit year in favor of its full four-digit format. Furthermore, the use of either a hyphen or forward slash should also be indicated, again simply to offer a uniform feel to information representation.

Time Standards

Naturally, the first question that needs to be answered is whether to use the 12- or 24-hour clock to represent time information. In the case of the 12-hour clock, a time in either the morning or afternoon is indicated by the use of the common abbreviations AM or PM, usually following the time. Standards usually dictate that the granularity of time information should not exceed hours and minutes, except in technical documentation where more accuracy is required.

A discrete piece of time information, such as the start of a meeting, is usually written as 1:30 PM (12-hour clock) or 13:30 (24-hour clock). Quantities of time may be represented by 2h45m (2 hours and 45 minutes), and ranges as 13–13:30 (for a meeting lasting 30 minutes). Where an increased level of detail (or granularity) is required, SI conventions should be followed.

Version Standards

Unlike date and time standards, version information does not seem to follow such widely accepted and rigorous formatting conventions, except where software version numbers are concerned. Here, it is widely accepted that the format should be major.minor. Therefore, a piece of software in its first release will be numbered 1.0, with the first minor modification being denoted by a version of the same software carrying the number 1.01.

Similarly, it has long been accepted that major version numbers indicate the addition of functionality, and minor version numbers simply show that various inconsistencies within the software (bugs) have been repaired. Any version before the arbitrary 1.0 level is often called "beta," although the simple addition of the letter *b* after the version number can also indicate this (as in 1.01b).

The actual version number can be mixed with the release date (formatted according to the target organization's standards) to give additional information to the end user.

Other Measurements

Besides measuring date, time, and version histories there will also be occasion to measure other values such as distance (measurements), area, or metrics representing the productivity of staff members. We will cover specific coding measurements in other chapters of this book; however, it is worth noting some conventions for indicating sizes of values are already widely accepted.

First, the target organization needs to be sure that everyone is using the same basic units of measurement, such as choosing British Imperial (inches, feet, miles) or SI units (centimeters, meters, kilometers). In addition, groups of staff members such as graphic artists usually have their own particular units, like DPI (dots per inch) or pels (pixels per meter), which need to be respected. There is little harm in choosing to mix units where the problem domain dictates, but such exceptions should be noted in the style guide developed and published by the target organization.

Staff Initials, Usernames, and E-Mail Addresses

In cases where the target organization does not reference users by number (archaic and impersonal), it is necessary to reduce their name for practical reasons. Smaller organizations will be able to simply adhere to the common three-initial rule. The author, for example, has always been referred to as GLT, although this could equally have been GWL, or even GWT, if required.

Larger organizations will of course run into problems where staff members have names that closely resemble each other, leading to the adoption of a two-three-four convention. Thus, the author could be referred to as GL (or GT), GWL (GWT), or the full four-letter initial GWLT. Hierarchical conventions may also be applied. For example, those higher up in the structure of the company may be referred to by more or less initials, reducing the probability of initial overlap and conveying status information in a simple manner.

Besides referring to staff members by initials, they will also probably require usernames for access to their computers on the target organization's network. Again, a common convention requires taking the first x letters of the staff member's surname, plus his first initial. Using the author as an example, a six-one convention would lead to *leckytg*. More common, however, seems to be an eight-one convention. The exact convention used is not important, so long as everyone is aware that a convention exists.

Finally, all staff members will probably need an electronic mail address. This can take the form of the username plus domain (as in *leckyth@company.com*), but this has drawbacks such as being a little difficult to remember for potential clients, and appearing less professional on business cards. The preferred standard is usually the entire name, as it appears on the staff member's business card, plus the company domain.

Spelling and Grammar

Most modern word processors have built-in spelling and grammar checking tools. Usually, these can be set to follow a specific standard, but can be altered by the user to reflect the working language of the document. Should the target organization decide to dictate the working language for all documents, then the word processors should be set up to reflect this decision. By a similar token, all staff members must be made aware of the decision to follow a specific language standard.

This is particularly important where variations of a language (such as U.S. vs. UK English) exist. It should be a conscious decision to choose between variations of a language to remove ambiguity through the corporate document set.

Similarly, grammar checkers can also follow the company ruling on language use; however, the majority of grammar checkers have proven to be ill equipped to deal with technical, legal, or specific language usage. Thus, their use is not recommended, except in cases where the software manufacturer specifically supports the style of language in use.

Document Writing Style

This is a book about software engineering, and so the style reflects the subject matter. Similarly, the writing style dictated by the guidelines of the target organization should be a reflection of the readership, which will change depending on the document being composed. Thus, several styles should be defined, ranging from memo styles, letter writing styles, and the all-important contract writing style.

To aid in the definition of style guidelines, a good reference manual such as the *Chicago Manual of Style* can be used to define the boundaries within which the documents should be structured, and the level of language used. The tone is also important, as it will either aid or hinder the transfer of information, which is the aim of documentation in general.

We will be covering specific kinds of documents later in the book, along with the style that they should use, but a general style should be adopted depending on the industry. In the games industry, a more dynamic tone might be appropriate, whereas clients being approached for financial software might appreciate a somewhat drier approach. In any case, the target organization needs to establish its own personality through the writing style guidelines that it establishes.

Currency and Value Representations

The currency convention might seem superfluous at first glance; however, it is necessary to decide which currencies are used in which circumstances. Although office supplies might be quoted and invoiced in the local currency, it may well be the case that clients in different geographical locations require work to be quoted in different currencies, with different levels of sales tax.

The accepted currencies should be listed, although this might be a dynamic list, along with the use for each currency. Where conversions are necessary, the rate source also needs to be established, and the same source for currency conversions used by all staff members when quoting for work toward clients, or requesting budget for items purchased outside the target organization's immediate geographical location.

Following currency conventions, there also needs to be a strict set of guidelines that details how values are to be represented, again to ensure that there is no ambiguity or misunderstanding. Commas, for example, may be used to separate orders of magnitude:

    1,000   10,000   100,000   1,000,000

or as a decimal separator:

    1,00   100,00

So, what happens when both are needed? Clearly:

    100,000,00

does not make any real sense. In which case:

    100,000.00

might be more realistic. Choosing the correct separator is a matter of convention and a logical decision, tied to the norms of the geographical location of the target organization, the currencies in use, and the problem domains addressed by the software being produced. In brief, a standard needs to be decided on, and passed on in the style guidelines.

Volume Organization and References

When deciding how sets of documentation are to be organized, and what naming convention is to be used, it pays to think about how they will end up being referenced. For example, it is well and good to organize documents in a hierarchical manner—this will ease the filing and management of the documents—but doing so can often lead to some strange references, such as V2D3XA. For internal documentation, this kind of reference is probably acceptable; what a client will make of it is an entirely different concern.

For those pieces of documentation destined for a third party, who may not be clued in to the exact operational and development environments of the target organization, titles in references are much more readable than esoteric number and letter abbreviations. For example, "Programmers Guide, Data Dictionary" is better than "D3XA," although the programmers may refer to it in this way once they become familiar with the convention.

However, even if we are to assume that the programmers will eventually become completely happy with referring to documents by their volume reference, the style guide should lay out some kind of convention for the preparation of specific document sets. The Functional Requirements, for example, is likely to be a multi-part document, if not an actual volume. It pays dividends to try to ensure that every Functional Requirements document is laid out in the same way, such that any *XA* refers to a glossary of some kind, for example.

Minuting Style

Projects tend to be made up of two distinct work areas: meetings and actual work. Each meeting is usually required to discuss some aspect of the actual work, and probably take a decision. Internal and external meetings should be treated with equal amounts of respect; in the same way that Chapter 1, "The Liaison Center," showed how internal and external clients were equally important.

Thus, writing minutes becomes an important part of the project documentation. All meetings need to be minuted, be they telephone conferences, progress updates, or more formal meetings. The minutes must then be circulated, commented on, and then agreed to (accepted) by all parties. Only one person should provide the initial minutes.

The way in which the information is presented and the detail of information contained will vary depending on the meeting subject, participants, and outcome. However, the style guides relating to staff initials, dates, times, and so forth must all be respected, and there should be additional guidelines indicating how the minutes should be approached.

Some organizations prefer the verbose to the terse. Some only require that decisions are noted, others that the entire discussion is transcribed (making recording the meeting a necessity), along with the feelings and personal notes of the person taking the minutes. The bare minimum will probably be the meeting title, location, date and time, a participant list, and agenda, all on the first page.

What follows is essentially a series of comments under each of the agenda points (agendas are discussed later in this chapter), with a brief note as to who made what comment. Depending on the style chosen by the target organization, the level of detail used to report the input of the participants will change, but the basic elements should remain the same.

As before, it is paramount that the style guide is respected across all minuted meetings, so that they all represent the meeting at the same level of detail. Since all participants will be required to sign off on the minutes, there is no danger of missing anything of great importance, since each participant will be trying to place his own view within the context of the minutes themselves.

One final point: Beware of verbose accounts of the proceedings. Reading the minutes of a meeting is almost as tiresome as writing them, and the fewer pages that need to be scan-read, the better. Minutes should also appear within 48 hours of the meeting taking place, where possible, making Friday afternoon the worst possible time for a three-hour progress update.

Agenda Style

Finally, all meetings need an agenda. This is to communicate the goal of the meeting to the participants, and is usually set by the meeting organizer. It should probably not exceed one page, preferably one side only. The contents will be, at a minimum, the title, location, date and time of the meeting, participant list (anticipated), and a series of points detailing the topics of conversation.

Each organization will probably have its own sets of agenda templates, depending on the meetings that are to take place. For example, a team progress report is likely to always contain broadly similar topics, with specific entries cropping up in cases where a particular problem has reared its head.

An "Introduction and Welcome" point is always good, especially if the participants do not know each other, and is followed by "Acceptance of Previous Minutes," in cases where the meeting is a recurring event. The meat of the meeting follows, and the topics will vary due to circumstances; however, the last two should always be "Any Other Business" and "Next Meeting." These last two give a chance for last-minute agenda points to be covered, as well as formally deciding whether another meeting on the same topics will be required.

Naturally, the Agenda will form part of the audit trail and project documentation as will the Minutes, and they should be collected, referenced, filed, and stored as with any other piece of project documentation.

PROJECT DOCUMENTATION

By and large, the exact mix of software development project documentation will be dictated by several factors. The first is whether the project is a new development, an integration project that ties together two pieces of legacy software (especially when merging two existing products or services), or the enhancement of an existing piece of software.

Second, the paradigm chosen by the company will also affect the exact documents that are required, since those projects that follow an incremental approach, including a phase of prototyping, will have a different document set from those following a more traditional model. Data-driven applications will also have a different document set from real-time embedded systems, for example.

The problem domain will also play a part in choosing what documents should be included, as will the end-user profile. Internal clients and external clients may also have different documentation requirements, and the level of client involvement will affect whether there are pieces of documentation specific to a given external organization that need to be included, for audit purposes, within the document set.

Finally, there is a difference between those pieces of documentation that are handed over to the client as deliverables, and those that are retained for use inside the target organization only, and are not distributed to a wider audience. The following discussion, therefore, is a global approach that can be used as is, or integrated with the existing guidelines offered by the target organization. It tries to be all things to all people, and as such will seem too heavy for some projects, and inevitably will be missing some parts for others.

Project Phases

We will assume that a project follows a set of phases, which have, as part of their deliverables, a document set that supports the completion of that phase. Generally speaking, a project performed for a third-party client will follow phases resembling the following:

Proposal: Usually a result of the client issuing a "Request for Proposals," this phase requires that the contractor details the solution, cost, and a time frame.

Planning: Feeding in from the Proposal phase, once the Proposal has been submitted, the contractor should begin to establish how he will service the proposal, regardless of whether it is accepted by the possible future client.

Execution: Once the project has been planned, and subject to the client accepting the terms of the proposal, the project is executed within the terms of the proposal document.

Completion and Maintenance: Once the project is completed internally, with a finished, and tested, product ready to be delivered to the client, it must be formally packaged, submitted, and maintained within the agreed terms.

This last phase could also be called *Handover* in cases where no actual maintenance agreement has been contractually arranged. This is a rarity, especially since the chances of the project delivering on all agreed functionality seems to be difficult to achieve within the budgets agreed by software contractors. Since the goal of this book is to ensure that this risk is minimized, we could claim that maintenance agreements are indeed unnecessary. However, there is always the possibility that the client will require enhancements to the software within a reasonable time frame from completion of the main project, under an extension to the main contract.

Bearing the previous phases in mind, we can now look at the kinds of documents that are going to be produced for each of the four phases that we have outlined. The categorization of each document is not rigid, since it is impossible to predict the exact timing and resource constraints of individual organizations. It is important to note that each document is a deliverable toward the client, and that the organization should indicate how each project's documentation set should be structured, resulting in a section of the organization's style guide.

Proposal Phase: Proposal Document

Generally speaking, the proposal reflects the request made by a third party, and needs to include an Introduction that details why the target organization is a good choice for servicing the contract, their strengths, and some basic financial data supporting their position in the marketplace and offering accountability of their financial position.

The bulk of the document reflects the approach to the problem domain that will be used, emphasizing, where possible, other successfully completed projects that used similar techniques. The structure of the "Request for Proposals" (RfP) document should be mapped to the solution. If, for example, the RfP is broken down into areas of desired functionality, then the solution guide should be, too.

Finally, a brief section detailing cost, licensing (where appropriate), ad-hoc extension prices, training, and expected time frame to completion should be outlined. Further to this, it might also be a good idea to break down the entire project into anticipated person hours of work. If the client does not require this, at least it can be used to back up the time frame plan, or adjust it as necessary, so it is a worthwhile exercise to perform.

One final detail is that a management summary should be attached to the Proposal, along with a cover letter that details the nature of the proposal, documents enclosed, and future steps.

Planning Phase: Detailed Plan

Following the breakdown of the project into anticipated person hours carried out in the Proposal phase, a detailed planning document needs to be written, which substantiates the overall plan, and indicates to the target organization where the resources need to be concentrated in order to achieve the goals promised in the Proposal.

It is also a good idea to build in a *double contingency*. This means that there is a certain contingency communicated to the client, in the proposal, to allow for changes of direction, new requirements, or human resource problems. The amount taken (which will be spread over the different parts of the resource plan) should then be doubled, and resources adjusted within the target organization accordingly.

A double contingency is a safeguard that ensures that toward the client, the time is built into the schedule, but within the target organization, resources are increased to ensure that the promised delivery dates are kept. It has no bearing on the eventual cost (the additional person hours are "free"), and in the best possible scenario, will result in completing the project *ahead* of time, rather than late.

Planning Phase: Contract

Usually a mixture of legalese and absurdly long sentences, the official contract needs to reflect every possible contingency, while protecting both parties. A standard contract can usually be established that covers most projects, and is the document that is signed, once the law professionals have deemed it acceptable.

While the contract will be binding, it will also be difficult to read. One has to make sure, however, for the well-being of both parties, that all deliverables, including documentation, are detailed properly, along with milestones and target dates. Penalties for missing milestones or targets should be clearly stated so there is no ambiguity.

Planning/Execution Phase: Requirements Definition

This document is the personification of the wishes of the client, whose signature appears on the contract, in a language that can be understood by both parties. By this, we mean that there is a minimum of technical terminology and professional vernacular. After all, both parties are going to sign off this document as embodying every possible function that the client will want to perform with the finished product.

Once signed, it is set in stone, except where the contract allows for modifications, and must therefore be complete, unambiguous, and easy to read. It need not be lengthy, just as long as is necessary so that the client agrees that the contractor has understood the goal of the project.

In the event of a dispute, *both parties* will be held to the contents of the Requirements Definition. As such, it will probably go through several iterations in an attempt to translate the RfP into a working document for the project.

Planning/Execution Phase: Functional Definition

Another document that should be accepted, if not contractually agreed, by both parties, the Functional Definition breaks down the entire system into pieces, each of which is designed to fulfill a specific task. The Requirements Definition might state that the spacecraft needs to be able to correct its own trajectory, and the Functional Definition will break this requirement down into subgoals (detect trajectory, check trajectory, fire rockets, etc.) that are designed to satisfy the requirement.

The reason why this document may not form part of the contract is that it moves toward the technical domain of the problem solution. As such, it may not be adequately understood by the client to form part of the overall agreement. However, if both parties are willing to invest the time to explain and understand, then there is no reason why it should not be used as a basis for agreement.

Execution Phase: Functional Specification

The Functional Specification is the design document, or blueprint, for the finished software application. It needs to be technical enough to provide the bridge between the Functional Definition and actual program code, from the point of view of the programmer. In addition, it should detail all the other nonprogramming requirements, such as data storage, operating system, and so forth that will be required to service the contract effectively.

Due to its technical nature, it will probably not be appropriate to include the Functional Specification in the set of documents that form a contractual agreement between the two parties. In certain circumstances, however, it may be desirable, especially if some work is being subcontracted, and both parties are technically competent to review the document.

Execution Phase: Acceptance Test Plan

Once the software is written, there will be a need to prove to the client that it fulfills their requirements, as laid out in the legally binding supporting documents outlined previously. This is distinct from the usual unit testing and other quality control methods that we will also be covering in this book, as it provides the only way for the client to formally accept that the contractor has fulfilled the terms of the contract.

As such, the document will also form a part of the contract, and should therefore be accepted formally by both parties, and again, signed off by their legal representatives. The language must therefore be nontechnical, although certain portions may need to contain data sets that by their very nature will be written in a technical manner.

Thus, the Acceptance Test Plan needs to be formulated as a result of the Requirements Definition document, and include parts of, or references to, the Requirements Specification. Therefore, it will need to be at least begun before the formal start of software development.

Planning/Execution/Completion Phase: User Guide

In brief, the User Guide enables the end user to lever the power of the software application to achieve the desired result. In addition, it provides the only instruction for installation, removal, and operation of the software package. Naturally, the client needs to be able to verify that it covers all the functions that appeared in the Requirements Definition, as tested during execution of the Test Plan.

As one might expect, signing off on this document constitutes formal acceptance of the entire software package, associated tools, and probably the implementation of the software at the site determined by the client in the contract. In sum, it is the solution to the problem laid out in the original RfP.

The document carries a lot of weight, and should be treated with an according amount of respect—something often forgotten by the developer in the haste to meet the deadline set forth in the original contract.

Completion Phase: Maintenance Contract

The last document that forms the complete root set is the guide for maintaining the product once it has been released into the production environment. Changes to the software, whether a reflection of the inevitable pace of change within the client's organization or as a result of finding small errors within the package itself, need to be catered for, budgeted, and priced accordingly.

Each of the previous documents needs to be written taking into account all the standards that have been defined as a result of the first section of this chapter. The organization should also be prepared to hand over a copy of the style guide to the proposed client so they can familiarize themselves with the documentation standards laid down by the organization.

CODING STANDARDS

Trying to persuade programmers to adhere to standards that impose restrictions on their own particular coding style is often akin to extracting water from a rock. Professional programmers also tend to have egos that fill the room they're working in, and do not take criticism easily. Establishing coding standards, in the target organization, is going to fail if it is done overnight. It needs to be folded into acceptable working practices over time—unless, of course, they are just joining the organization from an educational establishment, and have yet to create much of a style of their own.

Bearing all that in mind, several books are available (e.g., *Code Complete*, published by Microsoft Press) that deal with specific coding practices, and how they can be deployed within a corporate software production environment. While it is unlikely that every programmer will stick to every rule the books offer, some basic standards can be implemented that make the code easier to read, maintain, and implement, while watering down programmers' personalities a little, without seeming offensive to them.

One of the most frequently quoted coding styles available under an Open Source agreement is part of the Linux kernel source documentation; specifically, a document that covers the way in which coding styles should be applied by those programmers wanting to contribute to the Linux code base.

If developers are following the Open Source mentality of software creation, and plan to release their code to the public (under the terms of the license that is applied to the Open Source components that they have chosen to use), then this provides an excellent starting point for a coding style manual. It has the advantage of being widely applied, stems from plenty of programming experience in the field, and is reasonably brief and easy to follow.

A good description of this document can be found in the Kernel Korner part of the Linux Journal Web site (*www.linuxjournal.com/article.php?sid=5780*).

Comments

In a discussion of the various metrics that M Squared Technologies uses to measure code metrics (*http://msquaredtechnologies.com/m2rsm/docs/rsm_analysis.htm*) for their RSM software used for benchmarking and producing code metrics, they note that a ratio of 10-percent comments to code is a suitable minimum. They also point out that the quality is important, and that the comments should provide a reasonable summary of the code that they describe.

Three kinds of comment are very useful, and do not impose upon the programmer too much. Each function (method) should contain a comment that describes what it does, and what the inputs and outputs to the function are, along with any return value. These comments should be of sufficient quality to include in the official project documentation, and a tool like RSM can extract them in a variety of useful ways.

The second set of comments should reflect any modifications that have been made to the code since it was originally written, and must be identifiable using a code to identify the author, and reason for the modification. Finally, there should be a comment for each block of code that performs a separate step in solving the immediate problem.

There needs to be a set of keywords established by the organization that detail the type of comment that is being made, so that an automatic documentation package can extract them. Functions, for example, can have lines of comments that are labeled indicating input, output, or operations performed. Areas of code dealing with specific parts of the software should also be clearly marked to aid in the creation of programmer documentation such as data dictionaries.

Block Separators

Many keywords require that a set of statements be grouped by using separators to encapsulate the statements, or keep them together. Conditional statements, loops, and function block separators all adhere to these principles. Languages such as Modula-2 and Pascal use the keywords BEGIN and END to keep the statements together, while C and C++ use braces ({ and }).

Some style guides allow the programmer to place block separators on the same line as the keyword to which they refer; however, in the interests of clarity, it can make more sense to place them on lines by themselves. For example, if we take advantage of the C-style block separator placement rules to keep the number of lines to a minimum, we can end up with code such as:

```
void MyFunc ( int nRepetitions ) {
  int nCounter;
  for ( nCounter = 0; nCounter < nRepetitions; nCounter++ ) {
    // Do some interesting things
  } }
```

Indeed, C guidelines note that line breaks are entirely optional, which can lead to some particularly difficult to read code samples. The preceding code snippet can be rewritten in an easier to read fashion as:

```
void MyFunc ( int nRepetitions )
{
  int nCounter;
  for ( nCounter = 0; nCounter < nRepetitions; nCounter++ )
  {
    // Do some interesting things
  }
}
```

However, this does take up much more space, which is sometimes frowned upon by programmers. There is likely to be a trade-off between the two factors.

Function Size

To maintain readability, it is a good idea to set a limit on the number of lines of code that make up a function. Not to mention the fact that it has long been established that human short-term memory can contain around five pieces of information at any one time without inadvertently forgetting one of them. This, it would seem, is a better, more qualitative approach to limiting function size.

The Coding Standards might indicate, for example, that any time a function is in danger of containing more than five logical blocks, it has grown beyond the readers' ability to adequately comprehend, or validate. In addition, allowing blocks of code to nest by more than five blocks, or having more than five possible outcomes to a given operation will all detract from the readability of the code.

DATA COLLECTION STANDARDS

Part of the way in which the progress, efficiency, and overall productivity of staff members can be monitored is by collecting relevant pieces of data. These can be anything from working hours, lunch breaks, and coffee pauses, to counting the number of lines of code produced per hour worked; however, any data collected needs to be of sufficient quality to be used for drawing conclusions.

Besides the Big Brother style of corporate data collection that is used to monitor employees' behavior and productivity, the employees themselves might have cause to collect data in order to perform their job; programmers, for example, might decide that the performance of the code they have written is suspect, and want to measure it in some way.

Whatever the actual use of the data that needs to be collected, it should always be done in a way that is standardized across the organization, and respecting the guidelines for empirical representation set out in previous sections of this chapter. At its most obvious, a guideline might state that, when measuring time, guessing is not acceptable, but that a quartz-based timepiece has to be used, or that the system clock is to be initialized to a granularity that measures discrete processor counts.

Similarly, when counting lines of code, the organization might choose to exclude lines that only contain block separators, or comments, or even keywords. Referring to RSM again, M Squared Technologies has identified three code line counts:

LOC: Lines Of Code (all nonempty lines of code)

eLOC: Effective Lines of Code; all LOC that are not comments or block separators

lLOC: Logical Lines of Code; complete code statements

The authors of RSM state in their user guide that the eLOC metric most accurately represents the amount of processing work that the source code performs, and is the metric that software engineers naturally arrive at when asked to perform a similar estimation of work performed.

Clearly, the organization needs to establish which measurement they need to use in order to gauge staff member productivity or performance. While this is the most immediately obvious kind of data that needs to be collected, many more can be identified by the organization, and it is up to the guidelines of the organization to make clear how the data should be collected, and against what baseline it is measured.

REPORTING TEMPLATES

An integral part of ensuring that communication between all parties is carried out in an effective, timely, and productive manner is making it as easy as possible to perform the more mundane reporting tasks. For example, all organizations involved in providing consulting services will have a standard time sheet that all employees should fill out, and standard accounting documents that are used to calculate the cost of the project, so that the project manager can keep the costs under control and provide important information to the payroll department.

From earlier parts of this chapter, it seems that much of the style guide makes the task of reporting more difficult. After all, there are plenty of guidelines for effective language use and restrictions on style, all of which at first glance are not designed to make it easier for staff to write documents. Much of the standards, however, revolve around making sure that everything that needs to be in the reporting document is present, and that the key pieces are in a standard format.

This is where the templates come in. The virtual corporate library should include reporting templates for all aspects of project management and software engineering. They should be stored in a central place, and used by all staff. The staff should be instructed, when creating a document for the first time, to check whether there is already a template for the document, and if not, ask the relevant authority (librarian) to provide one.

Communication Documents

The basic document set includes templates for letters, faxes, e-mail, memos, and those pieces of documentation that we discussed previously for managing meetings. Each will be adorned with essential corporate information such as the title, logo, address, and contact information. They will also contain areas for inserting (possibly automatically) the version, and document reference numbers.

Software packages such as Microsoft® Word allow for the easy creation of templates that can be used as a basis to make documents. At the very least, skeleton documents should be prepared to cover the major document types. Most modern word processors allow the creation of fields within the document, which can aid in automatically generating document references.

Using the power of today's office suites of software, document references can be stored in a spreadsheet, which automatically numbers them, and the title and author/recipient information from the document can be fed back into the spreadsheet to provide a list of all documents. This list can then be exported, sorted, and generally monitored as staff create an auditable paper trail.

This requires some fairly technical linking together of documents, using the capability of most suites of software to integrate functionality, thus providing a method of sharing information between them.

Development Documents

All of the specification, definition, and general preparation documentation needs to have templates also, if only to provide a shell in which to place the relevant pieces of project information. We have already discussed what information should be present, in a very vague way, and we will revisit these documents later in the book.

Slightly more unusual is the possibility to use templates for coding as well. Most developers probably do not actually follow such a rigorous procedure, but it helps if templates for source code are established that include a copyright statement, licensing information, change history template, and author information—at the very least.

More advanced templates for use when creating classes, and implementing their various methods, as well as general utility and library code can be created. Most development environments, and word processors (which can be used to develop source code), allow for the use of macros that will facilitate the use of standard coding conventions for creating all the various pieces of code that will be needed to create the software application.

SUMMARY

Implementing the advice that has been put forward in this chapter will probably not make the reader many friends. Somehow, many staff members do not react well to being told how to write documentation, and that it comes before what programmers consider the fun part of their job.

However, having a set of corporate standards that are documented and enforced helps to ensure that the project is successfully completed. The aim of this book is to arm those involved in software engineering in a corporate environment with the tools to increase their success rate. Standards, documentation, and effective management are all key to achieving this goal.

Therefore, following the advice laid out in this chapter should lead to a visible increase in uniformity, productivity, and effectiveness, by itself. However, merely implementing a set of standards will probably not be enough to fully lever the potential of the organization; they are merely the foundation upon which the other principles are based.

3 Specifications

INTRODUCTION

Traditional Software Development Life Cycle (SDLC) paradigms usually involve one or more steps designed to translate what the end user needs into something that technical specialists can implement. This is particularly true when defining software that is custom built to match the needs of the client, who is usually not part of the organization performing the development.

Often, this translation of client requirements into documentation that forms the basis for the design of the software solution is neglected in favor of actually beginning the software development project itself. However, the root of many problems associated with late, incomplete, and overspent software projects can be traced back to neglected documentation in the early stages of the entire project.

The early documentation phases of a software development project are often given names such as "Requirement Specification" or "Requirement Definition," but the names themselves are not very important. What *is* important is that there is a phase that attempts to encapsulate the wishes of the end user in a document that can be agreed on by both parties, and a stage for turning this nontechnical document into a set of more technical specifications that may or may not be understood in detail by the client.

Certain measurements have been taken based on real projects that indicate errors that have not been isolated in the definition and specification stages become more expensive to fix, sometimes on a scale of many orders of magnitude. Thus, it is very important to try to pin down the underlying documentation. It is worth spending more time on these phases than seems necessary, and holding back the start of the development part of the project until such a time as the supporting documentation is agreed to be complete, by all parties involved.

As a brief aside, it is worth noting that, in principle, each part of the software development chain could be performed by a different subcontractor. This is one reason why, as we describe the approach and theory behind creating specifications, we always make a point of separating what the system is supposed to achieve from how it is to achieve it.

In theory, at least, we could have different subcontractors to define the requirements of the client, specify the problem domain, define the solution, implement it, and then test it—five separate third parties in all. In practice, this will rarely be the case, but it is always worth putting in the extra effort to ensure that the documents are written as if this were the case.

INVOLVED PARTIES

It is worth taking a moment to try to understand which parties will be involved in a typical custom software project, since technically minded project teams often neglect certain aspects that later prove to be important. One such example, as we will see, is the misunderstanding that can often result from not understanding the point of view of a nontechnically minded end user.

The Client

Clients fall into several categories. First are those who present the problem and expect the developer to produce a solution. These are the easiest clients to work with, since they do not have any preconceptions and will likely adapt to the developer's way of approaching the problem.

The second category is those clients who already think that they know the solution to the problem. In other words, they believe that they know what they want. Software engineering courses taught often point out that these are difficult clients to work with, who can be resistant to being advised as to the correct solution in favor of asking for specific features they think will solve the problem.

Finally, there are those clients who see a technical solution from the start, and believe that they are simply hiring a team of programmers to realize the project on their behalf. These will be the most difficult clients to deal with, particularly if they believe that they are adequately qualified to suggest how the solution can be implemented, while not being technically competent to do so.

Even in cases where the client is perfectly competent to present the problem, solution, and expected technical implementation, the developer should be prepared to spend some time at the start of the project explaining that the solution will be of a higher quality if the developer is allowed to work through the problem individually.

This is not to bestow some kind of super arrogance on the developer, nor to belittle the technical competence of the client. It's simply a warning as to how things can go horribly wrong when the developer tries to fit the client's world view into his way of working, rather than the other way around.

Each developer will attack the problem in a different fashion. This is why, in hard real-time development (such as space shuttle control), several development teams are approached to create the final product. In the case of extremely important control systems, the resulting applications are run in parallel, and each decision they reach is fed into yet another system that works out where all the systems agree, and recommends the appropriate decision to take.

This way of working illustrates two things. First, all developers are different, and the client cannot simply assume that their way of solving the problem is the same as everybody else's. Second, even when the end result is supposed to be standardized, different systems may produce different results.

However, it is only natural that the client makes suggestions. They need to be involved so they understand why the developers made certain decisions. There is nothing worse than a client who is kept in the dark, and believes that they could have done a better job, should problems be found at a later stage in the development cycle. They may simply assume that the developer is incompetent, cancel the project, and move on.

Therefore, the specifications of the problem and suggested solution provide an insight into the decision process that leads to the final implementation. They provide a vehicle by which the client and developer can effectively communicate their wishes and come to a common understanding of the problem domain.

The Technical Staff

All projects need to rely on the technical opinion of the experts during the specification phase. These technical experts will probably create the final product, and need to be consulted during the creation of the specifications. There is a temptation, however, of technical staff to underestimate the complexity of a solution, or overestimate their own capabilities.

The reason for this is twofold. First, they might not want to appear less competent than their peers. There is usually a healthy element of competition among hardcore programmers, which can lead to good, solid code. Many organizations tend to encourage competition for this reason, but need to be aware that doing so can have a negative impact on the accuracy of resource estimates as programmers attempt to exaggerate their competences.

The second reason is that they can feel persecuted, as if they are always at fault when a project runs over time, over budget, and is delivered with errors. A cynic might add that they have good reason for this, since they are ultimately responsible for the quality of the final product. However, it is equally true that poor specification, and poor quality control, could also be blamed for a project with problems.

This is why the opinion of the technical staff is vital. It prevents the management from making promises to the client that the programmers cannot fulfill, and ensures that the programmers have a hand in defining the solution to the problem that the client presents. Therefore, they can be confident that the specifications match what they can actually deliver.

Technical staff often have the advantage of an outside view. When they first become involved with a project, it will be without the initial account management contact that the developer has already had through the sales team. Therefore, they will have no preconceptions as to the problem domain.

As such, they might recognize problems in the specifications and in the description of both the problem and proposed solution that would make it difficult to realize an effective implementation.

Part of this is also coupled with the education and natural problem-solving disposition of programmers and software designers. The specifications, then, provide a way for technical staff to be appraised of the problem and proposed solution, while giving them the possibility to correct and refine the documents accordingly.

The Management

The project management can be seen as the internal client, as they need to be satisfied in the first instance that the final product matches the wishes of their client. This puts them in a slightly peculiar position, since they need to be sure that they have really understood what the client is looking for, not only so they can try to win the bid, which will lead to the project, but also that they can be sure that the project team delivers on that promise.

Therefore, the management needs to be sure that the specifications are correct, constantly checking that the client is aware that what they want may not actually have been effectively communicated to the project team. The specifications really need to reflect the agreement between the client and the management as to what the system needs to be able to do.

This will then naturally lead to a set of documents that state how the system is going to perform the tasks that will satisfy the specifications. If the specifications are incorrect, either because the client has not communicated effectively with the management or the management has not communicated effectively with the project team, then it will be difficult to achieve a positive result.

Therefore, the management is the bridge between the developer and the client, ensuring that the specifications represent a system that solves the client's problem, and can be delivered by the project team within budget, on time, and following the specifications. Of course, they should not be afraid to reduce the scope of the system to ensure that this is so—even if it means possibly not being awarded the contract, which is better than having to deliver on false hopes.

The End User

In many cases, the end user will be a part of the client's organization, although there will also be occasions where the end user is an outsourced service provider used by the client to provide technical services to them. However, end users should always be treated as a separate entity from the procurer, whom we generally refer to in this book as the client.

The end users will use the final system on a regular basis, so they must be sure that the specifications identify all areas of the final system that will make their jobs easier, or even achievable. It will not always be necessary to include their point of view in all aspects of the specifications, since the system will likely also consist of parts that will not affect the job of the end user.

However, in areas such as user interface, input from the end users will be invaluable in ensuring that the specifications lead to a system with which they can work. There are those who will say that the specifications as agreed between the client and developer should not actually address the user interface, since that represents the "how" and not the "what" of what the system should be able to achieve.

However, ignoring the end user in the specification stage may lead to an underspecified system, which, once built, will become expensive to operate, and even more expensive to change in order to satisfy the requirements of the end user.

End users will typically be nontechnical and therefore unaware of the consequences of some of their requests. They will strive to communicate a Utopian view of the final system, in which all their needs will be catered for in the best way possible, regardless of the cost of implementing such a system.

Hence, their point of view will need to be filtered by the technical staff and proj'ect management such that when it appears in the final specifications it is closer to something that will improve the quality of life for the end users, rather than solve all of their immediate problems.

The Technical Writers

Finally, after all the meetings between the client and management, roundtable discussions between the end users, client, and developer, and the inevitable diagrams that will be drawn as part of the reporting process on the evolving system view, a meeting will have to be held with the people who are actually going to write the specifications themselves—the technical writers.

In fact, they will have to be involved from the very beginning. Trying to communicate the final shape of the system will require yet more checks to ensure that the technical writers have understood what the client and end user are expecting from the system within the bounds of feasibility. The issue of the price that management will need to charge to make the venture worthwhile must also be understood.

The technical writer has to try to convey the ideas of all parties to paper such that the specifications are concrete enough to form part of the contractual agreement between the client and developer. Hence, it will be a document of some legal weight, carrying the signature of all parties, written accordingly.

The effectiveness of the whole chain from client to technical writer is summed up by the accuracy with which the specifications match the vision of the client—not so much what they think they want, as what they actually need.

COMMON MISTAKES

The client can be singled out as the culprit in cases where there have been misunderstandings leading to inaccurate or incorrect specifications, which in turn lead to problems with the final implementation. The technical staff will claim that they have carried out the wishes of the client as laid out in the specifications, and the technical writers will claim that they accurately described both the problem domain and proposed solution.

It is important to realize that both the technical staff and the client have reason to feel that the blame should be shared. Unfortunately, the client will usually blame the incompetence of the developer, and the developer will usually claim that the client withheld information that might have made a difference in formulating the solution.

Since the client also agreed to the specifications with the developer, the developer might also have a valid complaint in that the client should have been able to recognize potential errors and misunderstandings when they read the final draft.

Ultimately, the client and developer both have very different views on the problem and solution, not to mention different skill sets. Since we, as software engineers, desire to offer a service to clients, the onus is on us to try to minimize the impact of any mistakes that might be made during the early stages of the project.

Technical Competence

People of a technical leaning often forget that other people may not share the same experience and competence as they do. They may also assume that they can discuss topics that fall outside their immediate area of expertise, and still be competent to make decisions in that area.

It is said that a truly wise person knows what he knows, and what he does not. Technical staff sometimes need to be reminded that there are things about which they need to seek advice.

This would not matter, except that these assumptions often find their way into the specifications because the technical person concerned made an estimation without finding an expert and asking him directly, and because he failed to communicate the fact that it is a best guess. Finding these errors requires some form of review of all the technical facts in the document. This is a time-consuming, but often necessary, task.

Peer review will help enormously in trying to ascertain where there are assumptions or estimations that have been presented as facts, in two key ways. First, if those involved in creating the specifications, or providing input to the process where specialized technical writers are creating the documents themselves, know that there will be a peer review process, they are more likely to check their facts.

Second, when a reviewer sees a fact that is presented without premise, and cannot trace the origin of that fact, he will call it into question. It may be that the person who has written the fact into the specifications knows where it came from, but that a third party would not.

This may not matter, except in cases where the reader has to try to guess at the origins of the fact, where doing so affects his understanding of the document. There is the danger that if the fact is based on a premise that differs from what the reader has assumed, his understanding of other related areas will be incorrect.

Hence, all technical specifications need to be checked for accuracy, consistency, and proofs, where necessary, and where such proofs aid in the understanding of the problem domain. Asking technical staff to back up every fact that they communicate with a reference to existing documentation is not usually necessary, depending on the system complexity.

Terminology

Many mistakes can potentially stem from misuse of technical terminology. Part of the problem is that the language spoken by all the parties involved may not be of the same technical level, and miscommunications often occur when two words carry different meanings for two people of different technical competences.

The word *database*, for example, might mean a simple Microsoft® Access created nonrelational database to the office staff, but a fully relational SQL-compliant database system, such as Oracle, to a programmer.

The difference may not seem that important to a third party observing the evolution of the specifications. However, an Access database can usually be installed on a standard workstation, and can be used by anyone with a little experience with Microsoft® Office products. Typical Oracle installations, however, may require a Unix (or Linux) server, dual processors, plenty of memory and hard drive space, and specialist knowledge for installation and maintenance. The choice of definition, therefore, in such cases, is important.

There are many more examples of this kind of misunderstanding, and one of the key criticisms that can be made of specifications is that the language used is too vague. The word *database* in the previous example would not be sufficient for use in the specifications of a system designed to manage payroll transactions, for example.

However, it can sometimes be obvious to technically minded people as to what kind of database, for example, is applicable given the system being specified, the expected performance of that system, and the resources available to it. This is one of those assumptions that we pointed out in the preceding section, where it might be stated that a database is required, but two readers might have different understandings of the capabilities of a database.

Without the evidence to indicate exactly what services the database will be able to provide, one reader might assume a different level of performance for the system than another. If the client assumes that the system is capable of providing adequate performance for their needs, but it turns out not to be the case, the chances of finding the error before the system is put into production is slim.

The worst-case scenario in this situation is that the system proves inadequate simply because there was a misunderstanding about the underlying capabilities of the supporting database. This could lead to redeveloping either the database or the software, once the entire system has been put into place—which could prove very expensive.

The way to remove any ambiguities that might be the result of misunderstanding a particular technical term is to include a glossary of terms with every document, or a global glossary for all project documents. Each time a staff member writes a phrase such as "the database," he needs to verify that it has been defined in the glossary.

If such a phrase indicates a part of the system that is being built to specifications, and is not a third-party or retail product, then the definition of its capability should also exist as part of the specifications.

User Interface Design

Technically minded people are traditionally not well equipped to design the user interface, which is the user-facing side of the system being created. The only clues as to what the system is capable of, from the user's point of view, are communicated via the user interface. Therefore, it is sensible to try to ensure that the key features of the system are exposed via, for example, the menu system, and not hidden in an obscure dialog box deep in the software application.

When programmers try to design user interfaces, they have a tendency to assume that the user has extensive knowledge of the underlying system. There are many examples of this in office automation history—from esoteric text editing software such as vi that has no user interface to speak of, through the slightly less difficult to use WordPerfect™ for DOS, which was driven by a contextual menu system, to today's fully graphical menu systems in applications such as Microsoft Word.

One might say that the main driving force behind this evolution in user interface design has been the change from text-only operating systems to ones highly graphical in nature. However, early versions of WordPerfect for Windows, for example, showed that although the underlying GUI elements were present in the operating system, the designers of WordPerfect failed to take advantage of them. There are more examples: GIMP (the freeware graphics processing package) as compared to Paint Shop Pro or Adobe Photoshop, among others.

These products are typically designed by technically advanced programmers and created with intimate knowledge of the underlying capabilities of the system. However, they are very difficult to use because the GUI reflects how the designers use the system, not how a typical end user might want to use the system.

This leads to a slightly strange situation in that the features asked for by the client and communicated correctly to the developer through the specifications might be present in the underlying system but difficult to get to, leading to the client being less than satisfied with the end product.

To avoid this, end users need to be involved in the specification of the user interface, and not just the technical development team. If they are, there will be no finger pointing in the final acceptance testing of the delivered product, and the project is much more likely to be completely successfully.

Some software engineering paradigms allow the user interface to be created with a rapid application development tool such as Microsoft Visual Basic®. The result is a prototype that can be used to ensure that the end user is satisfied with the proposed look and feel of the user interface.

Screen captures annotated with text can then be inserted into the specifications by way of communicating this look and feel to the software developers. Of course, it is a policy decision to make as to whether the user interface represents the implementation (the "how") or the specification of the solution (the "what").

Some software engineers might point out that the actual modeling of the user interface does not belong in the specifications at all, since it is part of the final implementation. However, it is clear that there might be substantial gains to be made for those systems with a high degree of interactivity to be created around a modeled user interface.

DIAGRAMS

Specifications need to contain both text and an adequate number of graphical representations that illustrate the principle requirements inside the problem domain, and their proposed solutions. We will look at the kinds of diagrams that should be present in the various documents that make up the specifications of a product, and how they can be used to remove ambiguity in the specifications.

Notation

Many technical schemes can be used to represent various parts of the system, and all have their place in the set of specifications that make up the system description. However, different documents will be destined for different audiences, and it is important to realize that those who have not had the benefit of formal technical training will not necessarily understand some of the symbols used in diagrams prepared by, for example, software or system architects.

Consequently, the notation used in each diagram should be chosen to reflect the anticipated audience, which may sometimes lead to diagrams that are less visually rich than the author intended. That's fine, as long as the concept being conveyed is done so in a way that is unambiguous and compact, and uses textual annotations rather than advanced pictorial representations.

Designing the Problem Area

One of the first diagrams to be created may not find its way into the formal specifications since it is designed to aid the understanding between client, developer, designers, and end users. From this general understanding, the formal specifications will be created, so it is an important, if informal, diagram.

The best way to construct the diagram is in the guise of a brainstorming session, with all parties mentioned in the preceding section present or at least represented. The reason for this is that they will each have a different way of looking at the system, and each should contribute his view of the problem area to the initial diagram that represents a collation of all the separate views.

If one draws a series of bubbles on a piece of paper, each with a specific aspect of the system in it, suggested by representatives of the various groups (end users, client, management, project team, and so on), it will quickly become apparent that there are many different ways to look at the problem, and solutions to it.

The result will be a reasonably chaotic collection of ideas, which will need to be organized into clusters of ideas that belong together. It is possible to short-circuit the process in the interest of trying to establish different areas of the system and reduce the need to spend a lot of time writing on white boards. The following groups of ideas crop up repeatedly:

Data: The abstract objects in the system.

User: Representation of different user groups.

Function: Required features and processes within the system.

Commercial: Concerns such as price, size, and equipment required.

Service: Maintenance, diagnostics, etc.

If a sheet of paper is prepared to contain these five areas, perhaps in five columns, it will be faster to fill the various categories. However, this level of prompting might lead to undesirable effects, such as individuals trying to suggest ideas that are outside their own areas of expertise. In addition, those who are responsible for a given area might try to fabricate items so their column seems fuller, or worse, ignore key problems in the interest of having fewer entries in their column.

The idea of a more freehand approach is that, since nobody knows what will happen to the collection of ideas, those involved will be more likely to come up with anything that is on their minds, within their area of expertise, without the element of competition that can creep in by trying to streamline the process.

There are some loose rules, or principles, that should be adhered to when creating the list of viewpoints. There will need to be a member of the brainstorming session, with a good grasp of viewpoint analysis, who decides what a valid viewpoint is, and is not, using these principles.

Viewpoints represent functions, or collections of functions, that are a feature or service that the system needs to offer. Each function must be completely contained within a viewpoint. Any information that flows around the system must do so between two viewpoints, such that only viewpoints can provide a source of input information, or a point at which information leaves the system.

As such, each viewpoint that information visits should perform some kind of operation on that data. In fact, we can take this a step further and say that the purpose of each functional viewpoint is to perform some kind of operation on the information that exists within the system.

This is different from the nonfunctional viewpoints that usually represent the boundary conditions, or constraints, of the system, and can include items such as cost, performance, and resource requirements. Each of these can be considered a separate viewpoint, but should be grouped under the appropriate heading.

With the various contributions in hand, they need to be organized further into those parts of the system that define what it is supposed to do, those that define other systems with which it will need to interact, and the abstract concepts, such as price, that identify the constraints within which the system has to be built.

In the *Procedures for the 5th International Workshop on Software Specification and Design*, [Finklestein&Fuks89] suggest a name for this—viewpoint analysis. They begin in the same way, and take the process further by creating a structured diagram for each of the viewpoints that are proposed by the people involved in creating the initial description of the problem domain.

[Somerville92] suggests that a mixed object- and function-oriented approach is appropriate when taking this last step, known as *viewpoint structuring*. The aim is to create a diagram in which all the viewpoints proposed are linked together in a hierarchical diagram of different levels of related functionality. Figure 3.1 shows part of the viewpoint structure diagram for a simple accounting system.

Constructing the diagram serves two purposes: to have a working pictorial description of the system that needs to be constructed, and as an input to the creation of diagrams that describe exactly what functions the system needs to support. The act of creating the diagram also helps to provide input to the creation of these subsidiary diagrams, which will make up the bulk of the specifications.

Large systems, potentially containing hundreds of viewpoints, will need to be spread across multiple diagrams. It will not be possible to fit all the viewpoints on a single page, either because of the physical size of the paper, or simply because the resulting diagram cannot be absorbed or understood by the reader.

It is much more effective to layer the diagrams such that they only show a collection of viewpoints that are contextually related. During the structuring process that leads to the diagrams such as that in Figure 3.1, each collection of viewpoints will be grouped according to the level at which they appear in the diagram.

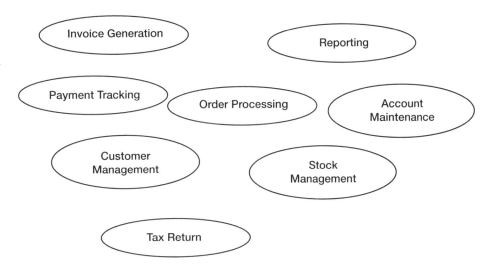

FIGURE 3.1   Viewpoint structure for a basic accounting system.

This means that the higher up the diagram the viewpoint appears, the more likely it is to be an object within the system, and the lower down the viewpoint appears, the more it will tend toward an actual function of the system. In performing this analysis, it may become necessary to split up viewpoints so that the diagram can be constructed in the most efficient way possible, in terms of communicating the required information.

Data Design Diagrams

At the base of the system will be the information that the system is designed to manipulate, and part of the system specification needs to deal with a complete description of this data. This process is known as *data modeling*, and the result should be a set of diagrams that show the structure of the information that the system will need to process.

This is different from the diagrams that will show the actual design of the data within the software system, but the difference between the two is admittedly slight.

The aim of putting data models in the system specification is to try to impose restrictions on the problem domain, but when the actual design is done, the aim is to create actual data objects that can be implemented by programmers.

These data diagrams will be a result of looking at the various viewpoints that were collected under the heading of Data in the viewpoint analysis. The titles within the Data collection will be abstract concepts such as log files, databases, and documents. For example, if one is defining an accounting package, we might expect a document such as the Invoice to be specified as part of the data collection of viewpoints.

When we try to convert the abstract notion of an Invoice into something more concrete in the specifications, we should model it as having various attributes: customer, amount, tax, due date, description of products or services rendered, and so on.

The diagram that we construct for use in the specifications that form a part of the contract between the client and developer will show an Invoice object, and the various attributes that it will need, with the emphasis on the *what* and not the *how*.

Once the specifications are accepted, and the whole system moves into a less abstract design phase, these diagrams will be enhanced to show how the system is to store and manipulate the various objects and attributes.

The viewpoint technique that we looked at in the previous section was built around the premise that each functional viewpoint operated on system data in some way, but we did not actually define what that data was, or how it should be represented.

Therefore, the data design diagrams complement the viewpoint diagrams in a description of the entire system. Without the data design diagrams, the specification of the system is not complete; by themselves they do not offer enough information to build the final system.

Figure 3.2 shows a typical data design diagram for a customer object, which will be needed by the accounting system for which we constructed a viewpoint diagram in Figure 3.1.

We have chosen an easy to understand notation for the data design, showing that the object in question (customer) has a collection of attributes. The central object, the customer, is shown as a box, with further boxes connected to it by lines that indicate the attributes (name, address, telephone number).

These attributes are then broken down further, and the final pieces of data are represented by ovals connected to the attributes by lines. Therefore, the address attribute has several components that refine it—number, street, town, and zip code.

We could go into even further detail, by defining each of the address components; however, this would begin to take us into the "how" area of the design, and the point behind the specifications is that they define "what" it is that we are trying to achieve.

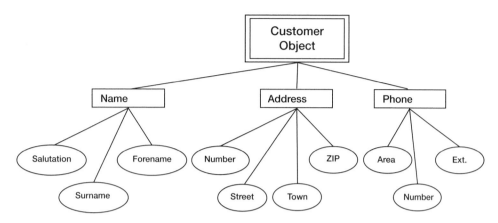

FIGURE 3.2   Data design for a customer object.

Process Flow Diagrams

Sometimes known as Action Diagrams [Somerville92], process flow diagrams show the various operations that the system is required to carry out, along with the data that will be required at each stage in order to perform the various functions required by the client.

The key to creating these diagrams is in isolating each functional viewpoint, and then expanding it to include the various operations, along with decisions as to what action to take depending on the outcome of each process, or action. This could be as simple as a binary possibility of success or failure, or it could result in a piece of data to be passed to another process.

This might, at first glance, be at odds with our viewpoint definition principles that we looked at earlier in the chapter. However, while each viewpoint must operate on a piece of data that belongs to the system, this does not necessarily mean that it needs to pass that data back out again.

The process behind isolating the processes that make up a specific viewpoint is called *functional decomposition*. We effectively push down on each process, and expand it into subprocesses that provide a graphical description of the functionality they are designed to provide. Figure 3.3 shows the first step in the functional decomposition of the simple accounting package that we have been using as an example.

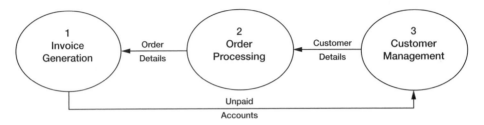

FIGURE 3.3 First order functional decomposition.

Figure 3.3 shows only the first level of the decomposition of the facilities that the system is designed to provide: Invoice Generation, Order Processing, and Customer Management. These are three key viewpoints in the functional domain.

Each has a data path associated with it, which shows how the information flows around the system. We call this kind of diagram a *data-flow diagram*, and it is more usually associated with the design of the solution, rather than the specification of the problem.

However, it is such a useful concept that we have borrowed it for use here since it should clarify the purpose of the system, and help the client to ensure that the eventual design will deliver software that matches their needs.

As we push down each process, we need to remember that we must stay away from areas such as GUI design, and interfaces with peripherals, such as printers. It is sufficient to state that, as an action, we want to print an invoice. The layout, how it is printed, what devices are compatible, and so on are part of the design and the system constraints. As such, they have no place in the specifications.

TIMING

As important as the specification of the requirements, definition of the problem area, and resulting formal functional specifications of the end product are, they are a small part of the overall software development life cycle, in terms of elapsed time. Indeed, the specifications may need to be flexible enough to allow changes to be introduced as the environment surrounding the system changes over time.

Phases

The traditional software development model can be broken down into very rough phases:

- Specification
- Creation
- Validation
- Maintenance

At each phase will be the necessity to create one or more sets of formal documentation (specifications) that define the exact nature of the tasks that must be performed to realize, or accomplish, that phase. Without an adequate description of exactly what needs to be done, the chance of the phase being successful is slim.

Indeed, in developing certain systems, the time required to perform the specification phase may exceed that needed in the creation phase by several times. It is worth taking the extra time so that the developer and client are in agreement as to what is about to be developed. The reader should pause to consider that if the specification phase is planned to require twice the elapsed time of the creation phase, this does not necessarily mean that there has been any wasted time.

If only half the time was used than had been allocated—thus squeezing the specification phase into a planned elapsed time that does not exceed that of the creation phase—the reader should be aware that this would probably turn out to be a false economy. There is a tendency in the industry to believe that the implementation phase should always require the bulk of the total planned time for the project to be completed, which often leads clients to try to save time by reducing other phases, most notably the specification and validation phases.

However, this will probably have a different effect, namely that by halving the specification time, one might double the creation time. In effect, the developers will have to do the same job twice, since the first attempt will probably result in a product that is not what the client wanted.

It is worth taking the time to get the specifications right.

The same applies to test specifications, which are part of the validation phase. If these are not complete, or inaccurate in some way, due to having not spent the time to ensure that they are correct, the creation phase may inadvertently also be extended as the final acceptance testing performed by the client reveals further faults in the system.

There is an inherent cost associated with finding and repairing any error in the system, and we mentioned in the Introduction that the earlier we find such an error, the less it will cost to repair. Figure 3.4 shows a graph that indicates the estimated cost of repairing errors against the time in the project at which they are found.

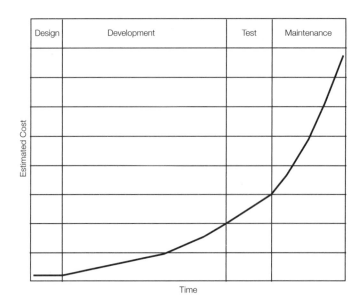

FIGURE 3.4   Estimated cost of fault removal.

While Figure 3.4 looks a little alarmist, it is arrived at by looking at the amount of work required to repair a fault in terms of what work has to be redone to fully fix the error. For example, if we find an error in the specification phase, all we need to do is update the text and possibly adjust a few references.

The same error, found during testing, requires that we fix the specifications, the design, and then redevelop that portion of the software, before retesting the affected area. Potentially, this represents a fourfold increase in the amount of work needed to fix the fault.

If we then take into account the fact that the interactions between the different pieces of the system increase exponentially as we progress through the SDLC, we arrive at the conclusion that, at each stage, there will be double the amount of areas affected by the same error in a given phase as in the previous phase.

Therefore, if we catch the error in the design phase, we will probably need to repair the error, and then retroactively fix something else as a result. If we only locate it in the development phase, we will probably need to fix other things too, say four of them, which will in turn have repercussions in the design, and specification phase. Finally, to locate something during testing is to start a sequence of events that could end in needing to validate an entire area of the system, which will cost much more than if we had found the error in the specification phase.

Of course, should we only find the error in the maintenance phase, once the software has been fully deployed, it becomes impossible to even guess at what the impact will be. The point is that it is well worth the effort to make sure that the system is as well specified as it can be before the design phase starts, because it is here that fixing errors will be the cheapest.

SUMMARY

Specifications are the bedrock upon which the foundation of the software project is to be built. Therefore, everybody needs to be involved: the client, so that they have a chance to put forward their requirements; the developer and technical team, so that they can comment on these requirements and try to better understand what the client actually needs; and the end users, so that they can see if the resulting system is something with which they will be able to work.

Producing specifications should not be viewed as a potential time sink. The time spent talking through the various aspects of the system, preparing documents that detail the most important parts, and agreeing on the wording of those documents, creating diagrams to enhance them, and generally trying to arrive at the same view of what the system is supposed to do is time well spent.

Software developers are fond of saying that an error in understanding costs one cent to fix on paper, but about a million dollars to fix in a live system. The key to avoiding these million-dollar problems is in being very aware as to what the system needs to be capable of with respect to the wishes of the client, and being able to effectively communicate this vision to those responsible for delivery of the system. This is performed by creating specifications.

REFERENCES

[Finklestein&Fuks89] Finklestien, A., Fuks, S. "Milti-party Specification," *Procedures for the 5th International Workshop on Software Specification and Design,* Pittsburgh, PA. 1989.

[Somerville92] Somerville, Ian. *Software Engineering,* 4th edition, Addison-Wesley, 1992, p. 73.

4 Product Development

INTRODUCTION

The first three chapters of this part of the book dealt with establishing a framework for the communication and management activities that need to take place in order to arrive at a point when the development of the product can begin. We also looked at the way in which the product should be specified and what controls should be in place that attempt to guide the development team into producing a product that is naturally of high quality, the assumption being that quality is expensive to force on a product once the development phase is over.

Before the actual software creation process can begin, we assume that a good quality set of specifications exists, and that the client has agreed that these documents do indeed describe the system they want to be built. Without this agreement, the development team has no guarantee that they will not have to rebuild pieces of the code because somebody misinterpreted the clients requirements early in the entire process.

To clarify, the more time that is spent getting the team cohesive, establishing a good pattern of communications, and a near-perfect specification document will ensure that less time is spent coding the product, largely because the first implementation is of such high quality. Some coding teams will approach the problem in an "implement and fix" approach, eager to get on with the coding, even if the specifications are incomplete, or not signed off, and then have to go back and "repair" the code. This means that the code has been reworked at least once before it is even tested.

Avoiding this will reduce the code, and ensure that any errors are caught in the early phases of creating the product—even before actual programming has begun. Sometimes, it is a very difficult to persuade the client that it is cheaper to spend time early in the project, not making much visible progress, than to begin writing the software straight away. It is up to the project manager to ensure that the client understands that industry sanctioned statistics show that the cost of error removal rises exponentially the closer the product is to final delivery.

PRODUCT DEVELOPMENT

The building, assembly, or programming phase of any project is the final chance to get it right; however, building software is not like building a house or assembling a television. It is so difficult to go back and alter a house once it is standing, or recall thousands of television sets because they have a design fault, that much of the process is taken up with ensuring that there are no faults in the specifications (architects plans, circuit design) that will result in the product not meeting the client's expectations.

These types of products also reuse bits and pieces of previous products, tried-and-tested technology, and third-party creations (such as bricks). All of the manufacturing procedures for creating such products are equally applicable to creating software, a point that most developers forget.

The reason why they forget, or choose to ignore this facet of software engineering, lies in the fact that it is only now reaching a point at which some of the components that can be used to create new products, coupled with the possibility of backtracking and altering the product once it has been delivered.

On the one hand, this is a powerful mechanism for creating prototypes, reengineering the product as the client's needs change, or just tweaking the user interface so that the end product meets the client's expectations perfectly. Indeed, there are paradigms for software engineering that rely on the ability to incrementally create the final product.

On the other hand, it should not be chosen simply because it results in visible progress being made. It may well be that the client is pleased when the developer can show progress, but attempting to fit other products into a new development, especially in a very specialized field, can sometimes lead to errors creeping in that are a result of a bad fit between the specifications and available components.

Choosing an Appropriate Paradigm

Rapid prototyping, used to create an initial version of a product, is a very useful technique when the product is mainly visual, resting on a framework of established components—not to mention its use for creating the graphical user interface (GUI) that the end user can employ to control the product. It is not acceptable when building real-time systems that control spacecraft.

Obviously, different paradigms will be used under different circumstances, but this book chooses to try to situate itself in a natural stance somewhere in the middle. If you want to build a hard real-time system, then the controls need to be slightly more rigid than those presented here. If a GUI-based system for collecting data (such as Web questionnaire programs) is closer to the product style, then some of the points can be relaxed slightly.

In Chapter 3, "Specifications," we saw how the creation of high-quality specifications can be achieved, by ensuring that all of the interested parties are involved in the creation process, following a paradigm that treats the process in terms of discrete phases, specification, design, development, testing, and deployment.

In some cases, however, it may be desirable to proceed with the development in favor of postponing the specification phase. This usually applies to high technology or innovative projects where it is necessary to establish whether something is possible before designing a vehicle for its deployment.

In such cases, the development work that is done can be considered part of the formal specifications, which is an aspect of rapid prototyping that makes it very attractive for use in software development.

THE SOFTWARE DEVELOPMENT LIFE CYCLE

We will be referring to the Software Development Life Cycle (SDLC) frequently in the middle part of the book, in which we deal with the software engineering part of the development mix. There are plenty of different views of how the SDLC should look, ranging from the classic "Waterfall" model, to more recent innovations such as the Spiral model proposed by [Boehm98], which attempts to factor risk analysis into the equation.

Generally speaking, methodologies have, in the past, been split into two camps—development and management. This means that we have one process model for doing the development work that leads to the final product, and another for managing the development process. Clearly, this is an inefficient way to develop any product, and the Spiral model is currently the only alternative that attempts to roll the software creation and management processes together.

Before we look at a workable, scalable variation of the Spiral model, we should analyze what it is about the Waterfall approach that has ensured that it is still one of the most widely used development paradigms for large-scale development projects. The answer is documentation; it is the only paradigm that lends itself to generating sufficient documentation to be able to gauge success and hence effectively manage the process, while remaining accessible to both technically minded and nontechnically minded personnel.

Briefly, the available options are:

Waterfall: A process of stages, in which the supporting documentation is signed off before the next stage can begin.

Exploratory programming: The final product is a result of trying various approaches.

Prototyping: Developing a well-specified prototype and adding functionality as appropriate.

Formal transformation: Producing a provable specification that is turned into a software product using a set of provable transformations.

Object reuse: Gluing a set of components together to create the final system.

The reality is that different paradigms are going to be useful in different situations. We are aiming at a Utopia in which we develop components using an augmented Waterfall model, and then use the components to create systems via a mixture of rapid prototyping and object reuse. This may sound complicated, but it is much easier to put into practice than it might first seem.

Exploratory programming and formal transformation methods are best left to experimental or cutting-edge projects, such as advanced AI or gaming applications. The three that remain all have their relative merits, and we will use prototyping and object reuse to augment the traditional Waterfall model to arrive at a paradigm that has the advantages of each, and minimizes the risks associated with them.

The Spiral model addresses this in a different way, by performing risk analysis, producing prototypes, performing verification and validation frequently, but still leaving the bulk of the testing until the end of the project, where it will prove most expensive to fix.

Augmented Waterfall Model

The traditional iterative Waterfall model looks similar to Figure 4.1, which is adapted from a similar diagram in [Somerville92].

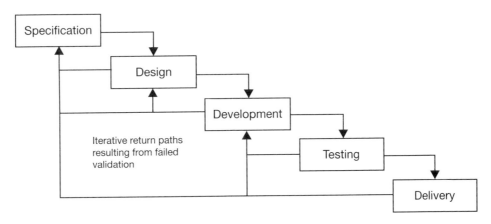

FIGURE 4.1 Iterative Waterfall process model.

We have chosen not to include the Requirements Analysis and Definition phase in our Waterfall model because it is not part of the software development process, but a prerequisite to the product development process itself. Return paths are included in Figure 4.1 that show logical recourse to modify the product at certain stages. It makes no sense, for example, to move from the Delivery phase back to the Testing phase if the client is not satisfied—at this point, it is only logical to either redevelop or redesign.

The problem with the Waterfall, and another reason why the Spiral model was proposed, is that it does not provide any indication of where verification and validation takes place, and the consequences of failure. The return paths in Figure 4.1 go some way to addressing this issue, but they do not provide a complete picture.

In Figure 4.2, we have added steps in the process for validation and verification, as well as introducing a phased approach, and the possibility to remove the Design and Development stage such that it may be replaced with something else, like component selection and integration, for use with the reuse paradigm of software development.

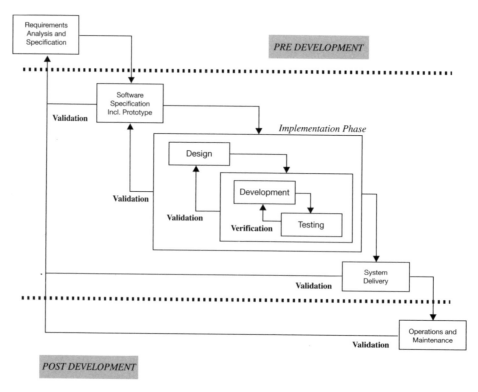

FIGURE 4.2 Augmented Waterfall model.

In fact, there is only one Verification return path—as a result of failure during component testing. All other failures are due to Validation exceptions. This is because each other phase will fail because we have not chosen to implement the correct product features, not because we have incorrectly implemented a correctly specified feature. The Augmented Waterfall model typifies the simplification of the process, generates enough documentation to satisfy project management requirements, and is flexible enough to embrace other, nontraditional approaches to software development.

We have dealt, or will be dealing, with the two areas that lie outside the scope of the actual development—Analysis, and Operations and Maintenance—elsewhere in this book. The remainder can be split into four key areas, which we will now discuss in some detail with reference to the underlying philosophies of the Augmented Waterfall approach to software development.

Specification

Assuming that the Requirements Analysis has been performed, and the Requirements Definition and Specification documents generated (see Chapter 6, "Requirements Definition," and Chapter 7, "Requirements Specification"), the first phase of developing the product that will satisfy the client's demands is to create a set of formal Specifications that will provide input to the implementation phase of the project, and allow the result to be validated correctly.

Of course, this relies on the fact that the Requirements have been correctly identified. Should this not be the case, then when the client sits down with the developer to validate the Specifications, they may conclude that there have been some misunderstandings that need to be corrected before the next phase can be attempted.

One of the key strengths of the Waterfall approach is that the client has to sign off documents that are generated as a result of a specific phase, so that the risk of passing into the next phase is minimized, as long as the client has well understood the documentation provided as evidence of good understanding. This mechanism, coupled with the iterative mechanism driving the Augmented Waterfall model, provides a similar safeguard to the risk analysis phases built into the Spiral model.

Part of the Specifications could be an initial product prototype, which should show the capabilities of the system without actually implementing any of the true functionality. This gives the developer a chance to try to apply some of the requirements to a real system, and gives the client a chance to catch any of the obvious errors in understanding before any programming work has been done. It is also much easier for the client to understand a prototype than some abstract definition of a product that might satisfy their requirements.

The idea of the Specifications is to provide an abstract description of what is to be built rather than how the various features are to be implemented in the final product. They are discussed in more detail in Chapter 8, "Functional Specification," but at a minimum, they need to contain a description of the information that is to be manipulated by the system, and the processes that are to manipulate the data, in black box terms.

Black box specifications give information as to the input data, and resulting output, or effect, but give no indication as to how the process achieves the desired result. Hence, the software is specified in terms of connected processes that can be validated by looking at the sum effect of these processes, which is appropriate given the intangibility of software products.

Once the Specifications have been accepted, the implementation phase can begin, which contains Design, Development, and Testing processes. These, as can be seen in Figure 4.2, are further grouped, which indicates the possibility to deviate from traditional development paradigms and apply others, such as object reuse.

Design

The design phase translates the Specifications into documents that provide descriptions of how the input data is transformed into either output data or some kind of effect. It is at this stage that we can either create a design that reflects each of the pieces of functionality contained within the Specifications, or one that contains designs for components that do not yet exist, and definitions of how components from the object library can be glued together to provide the required functionality.

This is one of the key reasons why we have spent so much time defining a good document repository and retrieval system—so that when we have a need for a certain component, and the specifications derived from clients' requirements, we can find one that has already been designed, developed, and tested.

The advantage is clear; every time we create a piece of software using object reuse, we create a product that has higher quality than if it was redeveloped from scratch. This returns benefits to the client, and helps to maintain the object library, as we might need to extend existing objects slightly in order to satisfy new requirements. If the changes to the object are correctly documented and the object checked back into the library, we also enrich the set of components available, providing future benefit to clients.

The reader will note that there is no failure return path for the Design process, but there is one for the entire implementation phase as a whole. This is where the Augmented Waterfall and the more traditional Waterfall and Spiral paradigms differ. Almost all the other paradigms offer Verification and Validation possibilities for the Design process. This begs the question, how do we actually Validate or Verify a design that might be incorrect?

The answer is almost a book in itself, and necessitates a detailed academic understanding of the various processes that is outside the scope of this discussion. Instead, we will assume that a failure in the validation of the Development and Testing process equates to an informal validation of the Design process: if the result of implementing the design is incorrect, then the design must be at fault. This is supported, as we will see, by the fact that the Development and Testing processes have their own relationship based on Verification of the code.

Hence, it is only after repeated failure of the client and developer to come to a validated implementation that the whole process is aborted and the Specifications reengineered. The assumption is that if the software that has been built does not appear to fulfill the client's requirements, there has been a fault introduced in the Specifications.

The Design process should yield a document that can be signed off, which indicates the mix of components that are required to make the software work. These are either completely new components or existing ones that can be used as is or after some form of enhancement. There is always a possibility that the client is unable to actually validate the design due to the technical nature of the document.

In such cases, they should be invited to validate the relationship between the prototype that has been agreed as part of the Specification phase and the components that have been isolated as necessary to complete the various tasks as required. Essentially, the client is being asked to reconfirm their validation of the prototype, and to state that they agree that if a certain set of components can achieve the desired result, then those are the components to use, in terms of input and output data or effect.

Development and Testing

The Development and Testing process is again contained within its own box in Figure 4.2. This indicates that it is another process that can be substituted for a nontraditional software implementation methodology—in this case, likely to be object reuse. It will probably be replaced with many instances, one for each of the components (or objects) that have been isolated in the design document, of which some will be reused, and others created from scratch.

Each piece of development that is performed needs to be tested, and the result of the test will mean that either the object or component can be added to the system, or that it needs to be altered in some way because the verification process has failed. As we mentioned before, those components that make up areas of the system that fail validation but have passed verification probably have flaws that date back to the design process, hence the return path in Figure 4.2 from the Development and Testing phase to the Design process.

In the Spiral model, reference is made to a test plan which is presumably signed off by the client, and which we have not made explicit reference to in the Augmented Waterfall model in the interests of simplicity. However, such information should be determined before the actual development is performed, and so the intention is to suggest that the test plan be part of the design documentation.

This also has the advantage that the client is probably best placed to know what the external boundary, exception, and normal operating conditions are likely to be, and hence can validate the test design much more easily than the developer. The Testing process is an attempt to verify that the software has been correctly implemented with respect to the design; again, this may or may not be what the client actually needs, but that is a design fault and not an implementation fault.

Delivery

The final part of the Development process is the Delivery phase. This is often overlooked, or only included as a milestone to be reached in an ever-changing project plan. In fact, it is a very important part of the entire development process if only because it is the first time that the developer hands a completed system to the client, along with all the documentation required to support it, and any training that might be required in order to use the system.

The Waterfall model as proposed by Somerville overlooks the Delivery phase entirely, with a phase for Integration and System Testing leading directly into Operation and Maintenance.

The Spiral model proposed by Boehm allows for an Acceptance Test phase that is closest to what we have chosen to call Delivery in the Augmented Waterfall model. The reader will note that there is a return path activated in case of Validation failure that leads from Delivery right back to the Requirements phase of the project.

The suggestion is that, after all the checks and balances that we have put in place to try to ensure that the final result will match exactly with what the client has asked for, should the process have failed to the extent that they are unable to accept the final product, then there must be something fundamentally wrong with the understanding of the client's problem in the first place.

It may not be desirable, or necessary, to work through the entire process again, but there may well be areas of functionality that have been badly misunderstood, and need to be reengineered. The exact consequences for the rest of the system will vary, but if the structured documentation guidelines have been followed, it should be fairly easy to establish exactly which parts of the system have been affected.

The process can then be followed for those parts of the system that have been identified as containing faults, and the new software system delivered once again to the client. Hopefully, few iterations will be needed before the client is willing to accept that the Development phase of the project is complete, and the system can be safely placed into live operation.

Hence, successful completion of the Delivery leads to Operation and Maintenance, which is concerned with the long-term life of the software system. There is also a return path to the Requirements phase, which deals explicitly with changes in the operating environment that will require alterations to the system that has been delivered.

By the lack of a similar return path to any other stage, we are indicating that the client expects the developer to be capable of delivering fault-free software—which should be the goal of all developers. This is not such an unlikely condition as it might at first seem. After all, we have allowed ample scope for discovering errors in the Augmented Waterfall model, and the software that is eventually put into production should, indeed, be fault free.

SUMMARY

The readers are encouraged to find their own Augmented Waterfall approach to software development. By using the Augmented Waterfall approach as specified here, many of the problems inherent in software design and development will be tackled in a way that is designed to ensure a high level of success.

This is not to say that the traditional Waterfall or Spiral models are not appropriate for software development, because they are. However, they do have issues, which means that they may not be a good fit with the philosophy of corporate software engineering that we present here.

The Spiral model, for example, may prove too overburdening for use in small-scale projects, as it contains many checks and balances that are only necessary in certain, extreme circumstances. The Waterfall methodology scales well in both directions, but provides no formal verification or validation, and does not adapt itself easily to the component reuse driven paradigm that object-oriented design and programming promotes.

Hence, the Augmented Waterfall model is a best fit for the circumstances—we need something that is scalable, allows verification and validation, promotes both the idea of prototyping and object reuse, and is easy to understand and apply.

There is also the question that comes up repeatedly in industry: How many times do we attempt to fix something before deciding that it is beyond repair? In other words, how often do we go through a cycle (such as the Develop—Test—Develop cycle) before we admit defeat and move back to the previous phase?

The answer is that there is no hard and fast rule, and that it will always be a question of feeling. As a good rule of thumb, however, one can take the initial estimate, and halve the allotted time for performing the process again at each verification or validation failure. When there is no time left, we assume that it will never be right, and throw it back to the preceding phase.

Thus, if we say that the development of component X will take four weeks, and verification fails, then the next attempt must be completed within two weeks. If it fails again, one week is allotted. If that fails, the component is abandoned and passed back to the design phase. By adhering to this principle, we can be sure that the project will not be extended indefinitely, and we have a good idea of how much extra time should be tacked on to the end of the project plan to allow for contingency.

REFERENCES

[Boehm98] Boehm, A. W. "A spiral model of software development and enhancement," *IEEE Computer,* 21 (5), 61–72, 1998.

[Somerville92] Somerville, Ian. *Software Engineering,* 4th edition, Addison-Wesley, 1992, p. 7–8.

5 Testing

In This Chapter

- Introduction
- Testing Procedures
- Test Result Documentation
- Automated Testing
- Test Data
- Storage
- Summary

INTRODUCTION

Left to its own devices, a typical programming team will probably eventually find itself in a loop that takes it from thinking about a problem, implementing a first solution, testing it, and then going back to recode pieces that do not quite work as planned. A proportion of these errors are a result of the way in which the entire project has been conceived, often related to an abstract data type, which requires that much of the modules dealing with that data type need to be reworked; an expensive error that should have been caught in the design phase.

A company called SPR has spent much time discovering what successful software companies classify as their development activities, and testing plays a large part. Requirements testing, design testing, function testing, unit testing, integration testing, and system testing are all mentioned. Some of these will be immediately familiar, others less so.

Broadly speaking, testing falls into the same category as documentation; programmers often resent spending too much time doing it, because they prefer to write code, and believe that they can go straight from a concept to actual computer code without writing anything down, and that it will work the first time. Of course, they probably will not admit to this, but every programmer is guilty of delivering untested code to the customer, often code that was not part of the original, "official" design documentation.

The approaches in this first part of the book are designed to compartmentalize the entire process, such that everybody, as far as possible, gets to spend as much time as possible on the activities that they enjoy, without sacrificing quality due to skipping, or being left out of, other activities that they might not enjoy but are needed to provide sufficient input to their principal job function.

Testing, it seems, usually falls quite late in the development cycle. It should start when the client first submits a Request for Proposals (RfP), in order to try to ascertain whether the logic presented by the prospective client or the team preparing the RfP response is adequately specified and correct. Errors, as we point out often, are much easier to catch if they do not exist in the finished product.

TESTING PROCEDURES

Before we can look at how to support testing, we need to determine what kinds of testing are required to ensure that the system has been correctly implemented. The purpose of the tests has also to be determined—is the testing taking place to ensure that the system is robust, or is it necessary to decide whether the system is capable of delivering the functionality that will solve the clients' problems?

The answer to this question will dictate whether the client is involved in the creation of the test data and philosophy driving the various scenarios that the developer needs to cater for.

Specification Testing

You can test the specifications in one of two ways: they can be tested against the client's perceived requirements, or tested for consistence with themselves. The latter testing can only be done if a formal specification exists; something we have not touched on in this book, because it requires that both the client and the developer have an in-depth understanding of a formal specification language, such as Z.

These languages allow the specification of the system in discrete terms, which effectively model the target system in a way that can be proven for correctness since each component is specified using mathematical semantics. Since these building blocks can be mathematically proven, the resulting system is guaranteed to be correctly specified, and, more importantly, unambiguous.

This does not help with testing the specifications against the client's requirements, unless the client is able to read the formal specifications. The only way to validate the specifications is by first validating the requirements, and then being sure that the transformation from requirements to specification has been correctly performed.

Thus, both the requirements and the specification need to be expressed in non-abstract terms; otherwise, it can be very difficult to visualize the proposed system. The developer needs to weigh the value of producing such robust and formal requirements and functional specification documents against the cost of their production, validation, and communication to the client.

Program Testing

Various types of test can be performed on implemented computer code, and while none of them deals effectively with conceptual errors relating to the design of the product, they will catch errors in the underlying implementation.

Unit testing: Individual functions are tested in isolation with a variety of inputs.

Module testing: Complete modules are tested in conjunction with likely and exception scenarios.

System testing: The entire system is tested for robustness in exception tests.

Integration testing: The system is tested within the context of likely operations and existing systems.

User acceptance testing: The end user or client is invited to examine the complete system.

One UK company has a policy that each line of code that has been written must be witnessed to have executed by the programmer responsible for that line of code. This is probably overkill in that programmers will have a tendency to find such an approach tedious enough that they will naturally shy away from it, not to mention the fact that just because a line of code executes does not mean that the system will work correctly, just that it will not halt during operation, whether it is behaving correctly or not.

Therefore, program testing exists to see if both the mapping of input to output data is correct—that is, the code performs the function that it was specified for—and that it performs well under error conditions, using accepted mechanisms for reporting that such a situation has been encountered. The interface testing, and mapping from a set of input data to a set of output data, needs to be performed in conjunction with a set of test data where the mapping is known and can be validated. We cover this later.

Unit/Module Test Procedures

In line with our philosophy for reuse at all levels, there should be one set of documents that deals with the procedures for testing specific implementations. Each module that is tested and has its quality affirmed will be placed at the disposal of any project requiring the functionality it offers. It will find itself being used in different situations, and test procedures need to be defined that determine how, why, and using what method testing is to occur.

Unit testing occurs on the smallest possible implementation unit, which means that each function needs to be validated in terms of data mapping, bounds checking, and exception processing. For example, imagine that we have specified that there needs to be a class for reading a file in a given format, and that it should report errors via an exception mechanism, then at some point there is likely to be a function to read a single piece of data from a file.

The test procedures need to specify that testing is required to test possible exception cases, such as files that are not open, files that do not exist, files that exist, can be opened, but are not in the correct format, and finally, files that can be opened, are in the correct format, but are corrupted. They also need to cater for cases where the file can be correctly read, but there is a problem with the internal representation that makes it impossible to store the read data in memory.

These are all exception cases, but we need to be sure that the implemented code, when executed correctly, leads to correct results. It is not sufficient to simply test all the exception cases, and assume that if they are handled properly, then the normal processing must, by inference, work. Therefore, the procedures must allow for ranges of correct cases to be tested as well.

The companion to Unit testing is Module testing, which needs to be performed to determine that, in the preceding example, the file reading class works to correctly parse an entire file of data and store it somewhere, as well as being able to clean up after itself when the data is no longer required.

When putting the procedures into place, it is safe to assume that the Unit testing has been correctly performed, and so we do not need to explicitly revisit cases that have already been tested as part of the Unit testing cycle. Taking the previous example, if we assume that we have validated the function to read in one piece of data, we can assume that, barring problems with internal (memory) storage, reading a series of pieces of data is going to work.

Therefore, we need only concentrate on testing the exposed interfaces of the class, such as the ability to read an entire file, because we know that the internal functions have been adequately tested. Of course, if the Unit testing procedures have not been followed, then the Module testing will fail, and could ultimately cause the downfall of the entire system.

System/Integration Test Procedures

Once all the modules (classes or objects) have been tested, complete portions of the system can be verified to work within the boundaries set by the environment. There is no need to go back and test all the exception cases, since we have done that during the Unit and Module testing, and we can concentrate more on the functionality of the entire system.

This does not mean that we can ignore the data validation functionality of the system, but that we do not need to check out-of-bounds exception cases such as memory corruption, file system errors, or input and output data mismatches, and can instead look at likely operating circumstances.

The System test procedures need to respect certain sets of test data, and certain sequences of events, and verify that if the system was to be running, it is capable of performing correctly. These will be covered later, but should be agreed upon with the client before development begins.

Integration testing is probably the first time that the entire system is put together and placed in an environment that approaches the eventual production system. The interfaces to all other external systems (databases, networks, print devices, etc.) need to be validated, and as such there will be very little scope for revisiting any of the key test data sets that will already have been looked at during Unit, Module, and System testing.

User Acceptance Test Procedures

Finally, once the developer is satisfied that the system is robust and meets the criteria agreed upon with the client, it can be handed over to the client for Acceptance testing. At this stage, it is the client's responsibility to validate the resulting system and ensure that it meets their requirements, as it is the last time they will be able to make changes before the final handover.

The procedures should make it clear that it is not the time for clients to submit requests for new functionality, but only for pointing out where there are errors in understanding. The system will be robust, having been verified at several points to ensure that it is, but it may not meet with the client's expectations, despite all the checks that we have put into place throughout the Requirements Specification, Functional Specification, and Development cycles.

Generic Test Procedure

The diagram in Figure 5.1 shows how the flow of control moves around the development system for a typical development project involving the creation of code from scratch. In cases where prebuilt objects are simply glued together to produce the final result, those parts in gray can be left out of the test process, as it is assumed that reused objects have already been tested fully. However, should they need to be altered before they are used in the context of the new project, they do need to be retested.

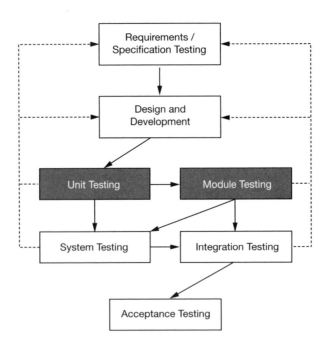

FIGURE 5.1   Test cycles.

The return paths in case of validation failure are shown with dashed lines. In some cases, it will be necessary for the code to be returned to the Requirements, Specification, or Design and Development phases of the project depending on what the exact problems were that caused the validation failure.

TEST RESULT DOCUMENTATION

The only way to make sure that the programmers follow test procedures is by defining specific documents to record the success of tests as well as guaranteeing a certain level of compliance with the specifications by using a series of compliance sheets. These sheets may also have been used during the RfP process to ascertain whether the proposed system is capable of fulfilling the client's requirements.

Reporting

Test reporting simply identifies that a given test has been carried out, when, and by whom, and whether the test was successful. The level of test reporting required will depend on which phase the tests are being conducted in, and by whom. Generally speaking, Unit and Module testing is performed by the actual programmers, while additional Module, System, and Integration testing will typically be carried out by the quality assurance department, with help during Integration testing from the client. Acceptance testing will be carried out by the client, with help from the end user.

Therefore, during Unit testing, the only document that will be issued will reflect the success of the tests, since the programmers will correct items that are not implemented satisfactorily as they go along. Module testing carried out by the head of the programming team responsible for that module will probably need to pass back test result documents to the programmer whose code produced the error, but it will probably not be more widely circulated than that, except during an audit.

The results of System testing will be passed from the QA team back to the development department, in case of error, so these must remain a formal part of the project documentation. If the client becomes involved during the Integration testing phase, the test reports need to be sufficiently well thought out to provide information sharing between the client, developer, and QA team.

Finally, the Acceptance Test Report needs, as its title suggests, to be a single document in its own right that isolates only those areas of the system that do not live up to the client's expectations, with reference to the original requirements. Part of this document will be a set of Compliance sheets that will help to isolate exactly where the problem occurs.

Compliance Sheets

A Compliance sheet is a special kind of reporting document that shows a specific test been carried out, and it has been validated against the original requirements of the client. Of course, it is understood that the client needs to be involved in the creation of Compliance sheets, and because they can exist at different levels, a hierarchy of Compliance sheets is entirely possible.

At the very highest level, the Compliance sheet should fit on a single page, and needs to contain a single line for each area of functionality that the client wants the system to exhibit. As each line is validated, it can be expanded to indicate what underlying Compliance sheets have contributed to reaching the specific goal that is listed.

These underlying Compliance sheets will have to be created at the same time as the design of the system, because it is at this stage that the various tasks are isolated and described sufficiently to know if the particular feature has been delivered. A completed top-level sheet indicates that the system has been correctly built, while the intermediate reporting ensures that the system is robust.

Therefore, by specifying and documenting tests via the use of direct reporting and Compliance sheets, we have satisfied two major components of quality control: robustness and fitness for use for a given purpose. This assumes, as always, that the purpose has been correctly isolated in the first place.

AUTOMATED TESTING

Before we look at specific test data sets that can be used to exercise a software system, we need to point out that many areas of testing can be automated. It is easy to forget this, and the developer needs to be prepared to make some investment in tools, procedures, and probably additional staff whose sole responsibility is to ensure that testing is correctly carried out.

The purpose of automating testing is to release the programming team from the requirement to spend valuable programming hours trying to verify that their code works in every conceivable situation, and some more that could not possibly be planned for. To do this, we need to use some clever techniques to simulate many hours of protracted use of the software system; both at the program level and user interface level.

Program Testing

Ideally, we would like to write a function and then be able to pass it to a program that would know about the function, the inputs and outputs that it is supposed to be able to deal with, and test all possible combinations of values and spot where potential problems exist. While this is probably impossible to implement for all software projects, it is reasonably easy to provide a generic test harness that can be used by the programmers to verify that their function fulfils the requirements for robustness.

Such a test harness needs to provide:

■ Extensive primitive testing (numbers, strings etc.)
■ Uniform exception handling and reporting to catch errors
■ Adequate result reporting

The test harness itself should be compiled into a separate program, which will run and generate a report that reflects the results of the tests carried out. This will become clearer with an example. Suppose we need to implement a function to copy one string into another:

```
int CopyString ( char * szSource, char * szTarget );
```

For those readers not familiar with the C language, the preceding line of code defines a function CopyString that takes two pieces of memory, of type character as arguments and returns an integer value, which can be used to indicate success. When dealing with strings in memory, there are a number of areas that we might like to test, including:

■ Invalid memory blocks
■ Destination smaller than source (memory overflow)
■ Invalid string data

We will expand this set when we look at the test data section of this chapter later. For now, let us just assume that we want to validate the CopyString function with reference to passing invalid memory blocks. As programmers, we might construct a test harness that looks akin to the following:

```
...
// Check for invalid source memory block
nReturn = CopyString ( NULL, szTarget );

if (nReturn != ERROR_CODE)
  printf("Error — NULL source string not caught\n");

// Check for invalid target memory block
nReturn = CopyString ( szSource, NULL );
```

```
if (nReturn != ERROR_CODE)
  printf("Error – NULL target string not caught\n");
...
```

Again, nonprogrammers are going to be at a slight disadvantage when looking at the preceding code, so we should explain that all the code does is call the function with an invalid argument, and then verify the return code. If the return code does not indicate that an error has occurred, then we report this fact.

So far, so good, and it does not look like a lot of code to have to write. However, having to do it for thousands of functions represents a significant amount of code. Each line could potentially contain errors, and would need to be verified, so it is much better to either generate test harness code, based on the function definition, or at least have a generic template that programmers can adapt to test their functions.

One step further from this is the requirement to verify ranges of possible, and exception case values. Remaining with the idea of verifying a string copy function, we need to be able to check that:

- Different sizes of strings are correctly handled (from minimum to maximum).
- All possible string values are correctly handled.
- The string copy function is correctly implemented.

To do this, we might like to be able to construct code that looks like:

```
...
// Check an incredibly large source memory block
nReturn = CopyString ( BIG_STRING, szTarget );

if (nReturn == ERROR_CODE)
printf("Error – Function returned an error\n");

// Check that the source block is the same
// as the target
if (CompareString ( BIG_STRING, szTarget ) != EQUAL ))
  printf("Error – Big string corrupted\n");
...
```

We are cheating slightly, because we assume that the CompareString function has been correctly implemented, and it would have to be verified before use. Since string handling is going to be an integral part of most programs, there is a good chance that this function will need to be part of the standard test suite anyway. All implementations of C include a function `strcmp` as part of the standard libraries that essentially performs this function.

Test harnesses of this nature can be constructed for all cases where we are not passing user-defined types as arguments. In cases where user-defined types are used, then we need to be able to verify them by creating a test library that is project specific, which we shall see later in the section *Test Data*.

GUI Testing

Testing the graphical user interface (GUI) of software packages used to be a case of putting a tester in front of the screen, and asking him to press many buttons, manually enter many data strings, and generally be a guinea pig for the system under construction. This is a very labor-intensive way to verify that the user interface operates correctly.

However, it validates the underlying system operation too, so it does have some value. It is worth bearing in mind that, as a general rule, if we replace an aspect of human activity with a piece of software, we will generally lose the specific advantages that having a human do the job brings—such as flexibility to respond to errors in a way that adds value to the process.

Nonetheless, many of the mundane testing activities can be replaced with a piece of software that can be programmed (scripted) to press buttons and enter data as if it were the user, and record what has happened in the system. The software should also be able to verify that the operation of the system is correct with respect to the buttons that it has pushed, or data that it has entered. To do so, some kind of result needs to be passed back from the underlying system to the GUI. Since human users need also to be informed about the status of their last action, this is likely to already be in place.

This approach to GUI testing still requires human interaction to create the test scripts, validate that they have been correctly carried out, and look at the results. The actual *doing* of the tests is where the advantage lies in using automated testing, as it becomes much less labor intensive, and hence less expensive to perform.

We can also create test data automatically, as we already noted in the preceding section—we need only know what types of input are required in the system. Actually testing the interface requires that a further piece of software is created that is capable of simulating the user. Commercial software exists that can be scripted to do this; for example, *WinRunner*™ for the Microsoft Windows operating system.

Command-Line Testing

Software and tools that are executed from the command line, with a variable number of arguments, can be tested much more easily by creating scripts or batch files that can exercise the software application. These scripts can be created manually or automatically, using the principles outlined in the following section for the creation of test data sets.

It is also necessary to verify any files that might have been created as a result of the software running, as well as any screen output.

TEST DATA

Test procedures will be made or broken by the choice of test data used to exercise the specific application area being tested. Several levels of test data can be created, from the most extensive that verify the behavior of the system in all possible cases, to the most abstract, designed to test the flow of control and logic of the system without paying special attention to the kind of data being supplied to the system.

In general, the smaller the piece of system being tested, the more detailed the test data needs to be. If we take date testing as an example, we might have a date object that needs to be verified. One set of tests could be to check that the date object can validate supplied dates successfully. We would start with basic numerical verification, such as supplying negative integers for day, month, and year data, and work our way up to verifying that it correctly identifies cases where the number of days for a given month supplied is not correct, and performing leap year calculations.

We could break down these kinds of test data as follows:

- Memory allocation (including array bounds checking)
- Data type validation (ranges of data, from maximum to minimum, and everything in between)
- Argument validation (for use with functions)
- Logic validation (verifying cause and effect relationships with a variety of data)

Following on from this, we can isolate cases when each of the test data set types can be used as in Table 5.1.

TABLE 5.1   Test Data Set Types by Phase

Type of Test Data/ Test Phase	Memory	Data Type	Argument	Logic
Unit Testing	X	X		
Module Testing		X	X	X
System Testing			X	X
Integration Testing				X
Acceptance Testing				X

The reason it is important to select the right kind of data for the situation is simply because testing is very resource intensive, and we do not have time to check every possible combination of values at each stage of development. If, however, we can be sure that we have tested correctly at each level of abstraction, from the data objects through to the various pieces of the final system, we can limit the total number of hours we will need to spend actually testing, while also being sure that a high level of quality is maintained.

Dates

Testing date data is such an important, and often neglected, area of software validation that all organizations involved in software application development need to have a policy for date testing that can be reused every time a project is created that manipulates date information. This was particularly highlighted in the move from the year 1999 to 2000, where many errors were found, stemming from the use of a two-digit year representation, and coupled with the fact that many engineers had chosen to use "00" as an error condition. Hence, when the software moved from 31-12-99 to 01-01-00, strange things began to happen.

Testing this proved difficult—without looking at every line of code, it was impossible to know exactly how the system would react. Instead, many organizations decided to create test cases to cover moving backward and forward across date ranges from 1998 through to 2010, covering everything from leap year calculations to day of week determination.

Operating systems were forced to move backward and forward in time, and known sets of test data were built that were designed to verify that the software was capable of working correctly both before and after the dawn of the new millennium. Of course, all the tests needed to be run after the changes had been made that fixed the errors that were found, to be sure that no new bugs had been introduced as a result of the testing process.

The best way to create the test data is by filling a document with values that relate to a given moment in time and by using a piece of software to generate the usual cases:

- Date thresholds (1999–2000; 2000–2100)
- 29 February in nonleap years

It may seem a little odd to think of dates that are 100 years in the future, but in some specialist applications, such as spacecraft, where a satellite or traveling spacecraft may need to last for a considerable length of time, or insurance, where policies can last many decades, it is important to ensure that the same errors that were made in the 1970s and 1980s are not repeated in the 2070s and 2080s.

Scalar Variable Limits

All programming languages use variables to store data in, and some of the usual kinds of storage types are:

- Character
- Integer
- Floating-point

Character data includes any one of the 255 possible characters from the ASCII set, or indeed, any one of the ANSI or extended character sets. Integers are whole numbers, possibly negative, and floating-point numbers are those with a decimal point, again both positive and negative.

These are known as *scalar values*. They have an upper and lower limit, which varies depending on the exact representation required by the software application. Scalar variables need to be treated with some degree of respect, since any of the values that they can possibly represent could be passed to the software. If a function takes, as a parameter, an 8-bit integer, which allows representation from 0 to 64k, or −32k to +32k, then the function must be prepared to accept values in this range, even if it is only to report that the value handed to it was invalid in its context.

For a particular project, test case data can be constructed that tests the lower and upper bounds of acceptable scalar data values, as well as the out of bounds values associated with a system that has possibly begun a downward spiral toward failure. This being the case, test data needs to be created for all possible combinations of information, during unit testing, to ensure that the unit processes correct data correctly, and reacts intelligently to data passed to it that falls outside its ability to cope with it.

Therefore, by way of example, let us assume that we have a function, in C that has a prototype:

```
int Difference ( int a, int b);
```

which is designed to calculate the difference between a and b. To be sure of the behavior of Difference in all possible cases, we would need to create a set of function calls to put every possible value of a and b into the function, and verify the return value. If we only wanted to test the boundary cases, of which there are:

- lowest a   lowest b
- highest a   lowest b
- lowest a   highest b
- highest a   highest b

then the test code will be reasonably straightforward. What happens, however, if we decide that we want to add the mid-range values to the data set? We now need to add some more test cases:

- lowest a   lowest b
- midrange a   lowest b
- highest a   lowest b
- lowest a   midrange b
- midrange a   midrange b
- highest a   midrange b
- lowest a   highest b
- midrange a   highest b
- highest a   highest b

Now we have nine test cases, instead of four. Astute readers will have worked out that to determine the number of test cases that are needed to test all combinations, we need to apply the following formula:

```
<scalar range>number of arguments
```

Applying this to our previous example, we realize that we will need $65,536^2$ test cases, which is about 4 billion—a tough assignment if you have to create them by hand, so the test data will need to be generated automatically. We can reduce the amount of actual test cases by applying some environmental knowledge; for example, since all the values are unsigned, a must always be less than b.

Once this has been done, we need to know how we can automatically test for the success of the function, since expecting a person to trawl through millions of results is unreasonable. Generally speaking, this will be a simple matter if this kind of testing is restricted to the lowest possible form of testing. The higher up the test ladder we go, though, the more difficult it will be to analyze the results.

The steps required to create scalar test data are, then:

1. Identify the set of all possible cases.
2. Use environmental (language and algorithmic) knowledge to pare the test set.
3. Create (automatically) the set of test data.
4. Decide how to validate the results.

From these steps, we should have a document of test case descriptions, along with some test case data, for entry into an automated system, or custom-built test harness.

Memory and Memory Corruption

Most software applications will need at some point to work with collections of scalar information, known as *arrays*. An array is a limited amount of memory set aside to store a finite number of items, all of the same type. Memory can also be used to store an almost infinite number of items, of differing types. This is why it is important to test memory handling. This comes in three main forms:

- Array bounds testing
- Dynamic memory allocation testing
- Storage/retrieval comparison testing

In the first form, we are simply validating that the software reacts properly if a function is asked to access an item in an array that is outside its known size. If, for example, we have a string of character values, which we know to be defined with a length of 255, the software must be able to react correctly if one part of the system attempts to put 256 characters into the array, or retrieve the 257th character, since these are known to be "out of bounds."

Testing array bounds is simply a variation on the scalar value checking we looked at in the previous section, so no further explanation is necessary.

Dynamic memory testing requires a little more thought, but is again just an extension of the scalar tests with which we are familiar. We need to be a little careful in trying to estimate at what point we should stop allocating memory—in case the system has begun to over-allocate due to an error occurring elsewhere—and testing for that, as well as the possibility that the memory could not be allocated.

The third form is another example of needing a huge set of test case values, since we need to be sure that any kind of data that can be stored by the system can be placed into memory, and retrieved correctly, and that memory corruption does not occur when the system is in operation. To do this, we need to create test data sets that store as many different objects as possible, and as many different kinds of those objects. Again, this will have to be done using an automated solution.

STORAGE

The author has had the experience of software that was aborting for no apparent reason during operation, and later found that it was doing so simply because it had run out of disk space. This meant that the subsystem that was supposed to be logging errors was unable to do so, and in trying to report that an error had occurred (no more disk space), told the system to stop because it could no longer report errors.

This sequence of events led the debugging team to (incorrectly) surmise that there was a problem stemming from the last known operation that was logged by the reporting subsystem. Thus, the investigation that ensued started on the wrong premise, and many person hours were spent looking at the wrong part of the system until it was pointed out that it was impossible to save a file on the system because there was no disk space left. The disk space was upgraded, and the error could no longer be reproduced.

At the other end of the spectrum, there is the case where the system runs on a piece of hardware that supports extremely large disks, and the operating system incorrectly reports the size, causing the software to behave erratically. This can happen on PCs where the BIOS has not been upgraded to allow it to reference disks greater than a certain size.

In fact, the root cause of this is linked to the scalar variable limits, and is one of data representation. If the BIOS only sets aside a small number of bytes to contain the drive information, it will be unable to address larger volumes except by some clever mathematics. Memory addressing under the Microsoft Windows operating system used to suffer from the same problem, which is why it was eventually broken down into two pieces: a segment and an offset.

These days, of course, we are more used to multi-gigabyte disk and memory storage possibilities, and so the problem is unlikely to crop up except when there is a requirement to interface with a legacy system, at which point special care needs to be taken, which is the key to good system testing.

SUMMARY

In this chapter, we looked at all manner of different test data sets that can be created when we need to validate a system, and the stages in development when they need to be carried out. We did not speak in any detail about the actual methodologies that can be used to apply these tests, and should briefly mention them now.

Traditionally, testing is spoken of as being one of:

- White box testing
- Gray box testing
- Black box testing
- Certification

White box means that the internals are exposed, and is an equivalent to unit testing. The tester knows exactly what the unit is supposed to do, and how it does it, and can therefore target the test cases to be sure that all the logical variations are catered for. Gray box testing treats the code as opaque, but the data as transparent—testers know what data goes in and comes out, but not how the transformation works in detail. Hence, they need to create larger test data sets because they do not have intimate knowledge of the internal system. This is akin to function testing.

Black box testing treats the system purely in terms of inputs and outputs. Testers need to be sure that all possible data sets are catered for, because they do not even know how the data is stored internally. They are almost looking at the system as an end user, although it is usual to conduct black box testing prior to integration testing to be sure that the output of one piece of the system, when provided as input to another, is correctly handled.

Certification is a special kind of testing that is performed at the end of the cycle—rather like acceptance testing—and treats the system as an entity that is tested only by what it exposes toward the end user. The purpose is to ensure that the system is ready to be placed into a production environment, and will react in an appropriate manner no matter what the situation.

Part II

Principles of Corporate Software Engineering

In this part of the book, we approach the problem of software engineering in a way that is both powerful enough for large-scale projects, and scalable such that it can be used for humble projects as well.

The reader will learn everything from how to define the product, specify it in unambiguous terms, develop the software, and finally test and deliver it. We have chosen the Object-Oriented paradigm since it most accurately reflects the natural thought processes of solving a real-world problem in an abstract manner.

Chapter 6, "Requirements Definition," begins with the tasks that must be accomplished before any other work may be done in order to produce a high-level document that describes the end product and what is expected of it. The potential pitfalls of natural language in defining a nontangible deliverable are highlighted, as well as ways in which to extract the notion of what the client requires from what they think they require.

Following on from this, Chapter 7, "Requirements Specification," deals with how to specify, again at a high level, how the requirements will be satisfied. This goes hand in hand with Chapter 8, "Functional Specification," which defines in unambiguous terms what it is that the developer will actually aim to deliver. In this chapter, besides defining how the document should be written, the time is taken to specify how the client can be involved, even at the technical levels, thus reducing the probability of failing to deliver what the client thought they specified.

In Chapter 9, "The Object-Oriented Paradigm," we discuss how Object Orientation can be used to produce a code base that is easy to maintain, read, and extend. This chapter details the advantages of the OOP, as well as languages and tools that support it.

Following on from this, in Chapter 10, "Reusable Code Guidelines," we discuss reusable code and how to build it into the strategy of the development cycle almost from the start. We also discuss how to choose the correct definition of Object that enables practical reuse to be applied in the future.

Chapter 11, "The Object and Component Archive," defines a scheme for facilitating the storage and reuse of these Objects, such as practical techniques for indexing and storing objects so that they may be made available across multiple projects.

Once the basics are in place, we look at Chapter 12, "Coding and Language Choice," indicating, with the currently available languages, the options to consider when choosing the development language, or for some projects, language mix.

Chapter 13, "The First Prototype," describes the steps that should be taken to implement the correct features in the prototype, so that the client has a first impression of the look and feel of the product, while reducing the initial investment.

Following on from this, Chapter 14, "Adding Functionality," is designed to illustrate how the various components can be glued together to provide the functionality that has been defined, and provides a discussion of the various object and unit testing philosophies that may be applied. Finally, Chapter 15, "Delivery," describes how the final delivery is to be made, and provides a checklist of things that should be examined before the product is considered to be fit for release.

6 Requirements Definition

INTRODUCTION

While Part I of this book dealt with the organizational aspects of software engineering, Part II aims to provide guidance for actually performing the tasks that result in the creation of a product, or delivery of a service, to the client.

The Requirements Definition document is the result of the first task to take place once the contract has been won. It aims to accurately define what the system needs to do to solve the problem presented by the client. Key inputs and outputs need to be agreed upon, and defined, as well as the results of the various processes that the system needs to perform such that it provides a useful function to the client.

The document needs to be agreed upon by both parties before work on the product can begin, and as such, it must be expressed in natural language—or at least a representation of the system that can be understood by both parties. This is important because without mutual agreement, forming the basis of a contract, it is impossible to gauge whether the developer has delivered that which the client requires.

There are two problems to address: first, that natural language is an ambiguous means of communication at best; and second, the concepts that we are trying to convey are abstract and define an intangible product. This intangibility is characterized by the fact that the end result cannot be touched or viewed in any conventional way, and exists only temporarily as a program inside the memory of a computer.

The document might define what is to be achieved, but does not address how these results are to be achieved. Selecting the correct level of detail is vital at this stage, since while the system must be broken down into smaller pieces to correctly define the requirements, if too much detail is rendered there is a danger that the definition will begin to approach the "how" and not just the "what" of the system.

SKELETON REQUIREMENTS DEFINITION DOCUMENT

Many different approaches can be taken to the Requirements Definition Document, but simplicity is likely to be the key. It is easy to become entrenched in a document that is too complicated and attempts to cover every contingency, but the document structure should be rich enough that a precise collection of requirements can be collected.

The language of the document needs to be understandable by both the software designers and the client, and, being essentially a reference document, must be an integral part of the project documentation, and lodged as such in the library. The document becomes part of the collective knowledge of the organization that we described in Chapter 1, "The Liaison Center."

[Somerville92] lists a possible structure for the document, which we reproduce here:

Introduction

The system model

System evolution

Functional requirements

Nonfunctional requirements

Glossary

He also places a number of items in appendixes to the Requirements Definition Document, such as the Requirements Specification, which are better kept as separate documents, and not merely appendixes to the Requirements Definition. The reason for this is that we are working around a model for creating software that relies heavily on correct documentation, and as such, although it creates more individual works to be managed, it is better to have a great many smaller documents rather than a few multipart ones.

Part of the reasoning for taking this approach is also that it is easier to target documents if they cover a very specific subject matter. This is particularly well illustrated by placing the Functional Requirements Specification as an appendix to the Requirements Definition. As [Somerville92] points out, the Definition is aimed at the client, and should be written with them in mind.

The Requirements Specification, however, needs to be created using more formal language, in a much more precise manner, and therefore needs to be written in such a way that it should not be open to misinterpretation by the client or developer.

There are two points to be made here. The first is that the two documents should be kept separate, since, while the first will be readily accessible by the client, the second may not even be written in a way that they can read easily. Hence, to avoid clouding the issue, the client should only be invited to read the Requirements Specification under the advice of a technically minded third party, in the event that the client is of a nontechnical leaning.

The second point is that the Requirements Specification can form the basis of the contractual agreement between the client and the developer. In this case, if the client is unable to understand the language used, it may be impossible for them to sign a contract for development without seeking the advice of a third party.

This can be used as an effective check if the two documents are kept separate—simply invite a suitable qualified third party to read the Requirements Specification and then describe what they have understood, in natural language, to the client. As long as the client is able to agree with what they say, the contract can be signed. If not, then perhaps there has been a problem during the requirements capture process.

[Somerville92] also lists some other "appendixes":

Nonfunctional requirements specification: If these are important enough to have an impact on the system, they are important enough to have their own document dedicated to them. However, in the interests of effective communication, we have chosen to integrate them with the Requirements Specification Document.

Hardware: Again, we have chosen to make this a part of the Requirements Specification Document, since it will likely be referenced in terms that are technical by nature.

Database requirements: Since these will include a data model, and relationships between both internal and external data that the system needs to manipulate, these have also been moved to the Requirements Specification Document.

As a footnote, unlike Somerville, we choose here to refer to a Requirements Specification Document, and a separate Functional Specification. The reasons for this are that, as we saw in the previous discussion, we have chosen to include some technical items in the Requirements Specification that he places as appendixes to the Requirements Definition, and some nonfunctional items besides.

The document structure we present here is more in tune with the concept of the Liaison Center communicating via documents with all involved parties in line with particular standards and guidelines. In brief, the complete skeleton Requirements Definition Document now looks like this:

Introduction: Places the need for the system in the context of the wider target organization (client).

The system model: Shows the relationship between system and environment, paying particular attention to where the environment and proposed system need to interface.

System evolution: States the way in which the environment is expected to change during both the lifetime of the product and the time span of the related project.

Functional requirements: What the system needs to do.

Nonfunctional requirements: Constraints, imposed by the operating environment.

Glossary: Summary of technical terms and explanations thereof.

Index: Reference into the document of keywords and phrases.

We place more emphasis on the Introduction, System Model, and System evolution than we do on the actual Requirements because if we situate the product correctly with respect to the target environment—essentially putting ourselves in the client's shoes—we will be much more likely to get the requirements part right. This is not to say that we are neglecting the requirements, merely that we, as software designers, understand them better than we do the less technical aspects of the system; so we need to spend more time to try and understand them.

REQUIREMENTS CAPTURE

The art of interviewing, collating information, and presenting it for approval is known as *requirements capture*. The work that has to be done to turn the results of the requirements capture exercise into a set of formal descriptions of what functionality the system needs to offer is called the *requirements analysis phase,* and feeds into the Requirements Definition Document.

System Context

The system context defines the place of the system to be developed within the operational and organizational structure of the target organization. All systems need to have a role, with interfaces with other pieces of the organization that the system is put into place to serve.

It is a logical context, and a result of how the client sees the software with reference to the existing systems and processes that have been put in place to solve the particular business problems with which the clients find themselves faced.

Unless the developer understands how the client sees their universe, and how the system fits into it, they will find it difficult to create a system that will provide the functions that the clients require. This includes knowing how the end users relate to the rest of the organization, and the business circumstances that will cause them to use the system.

Therefore, the first part of the requirements capture process is to find out what all the various related business processes are, and how the system fits into that world view.

Operating Environment

The system context gives the logical environment into which the system will be placed, and the operating environment looks at the actual physical location of the system within the organization. If it is to be a server-based application, the operating environment will be very different than if it is to be a desktop application employed throughout the organization, with one copy per user, while the end result may be logically very similar.

Indeed the advent of internetworking applications and the maturity of the client server model mean that understanding and defining the operating environment can be quite a difficult task. In the end, the best result will be if the client understands that, from a software point of view, the actual operating environment may be very different from the way it appears to the end user.

Real-time systems, or those that are required to be available on a 24/7 basis, will be particularly affected by the operating environment due to possible conflicts with other applications. For example, it is no good if a system monitoring the filtering of air on a spacecraft crashes one night because another system has used up all the hard drive space. This may be an extreme example, but it does illustrate the importance of understanding the interdependencies of different systems in the same operating environment.

End-User Services

A vital part of capturing the requirements is knowing what the system is supposed to do, and what services it needs to offer to the end user for it to be of use. This may seem obvious, but many times it is assumed that the system is defined in a black box way, by the way that it interfaces with the environment and the end users. Somehow, the exact functions that it is supposed to offer are implied by the way in which it is to be used.

This may be true for a hammer, but it is certainly not true for a piece of software. It is the responsibility of the developer to ensure that they have managed to write down every single service that the system is required to offer to the end user. Then, the client can read them and understand that, if all goes according to plan, this represents the entire capability of the system, and they should not expect anything more.

The approach to take is rather like designing a word processor. Everyone who has ever used one knows that it has to have an area into which words need to be typed. They also know that it should be able to print out on a printer, and have some basic editing functions such as search and replace, and perhaps a good spell-checking module. How many people would say that an absolute requirement is the horizontal scroll bar?

Most systems designers would overlook the scroll bar in favor of looking at detail into all the advanced services, such as the possibility to create Web sites at the click of a button, simply because it is a more interesting aspect. Once the software is in the field, and a user begins to type a report with the software and finds that it is missing a horizontal scroll bar just at the moment he needs it, it becomes very difficult to add one.

Supporting Services

Supporting services are those functions that are required by the system to ensure that it continues to run smoothly. They include things that are needed for day-to-day management of the system, and ancillary functions that are not part of the end-user interface, but are there in the event that some form of routine update or check needs to be performed.

This might include functions such as adding a new printer to the system or performing a backup of the data (or, in dire cases, a restore of the system data). They are functions that the client might not actually have thought of, but need to be asked if they should be included on the grounds that not including them might cost more later to put in retrospectively.

The way to approach the requirements capture for such items is to look at all the points at which the system interfaces with external hardware or software, and try to imagine whether these aspects will ever be changed, or what happens when they become compromised in some way.

The requirements will be different depending on a number of factors. For example, it should be obvious that the automatic housekeeping required for a spacecraft that is destined for another planet needs to include functions for managing the hard drive space. After all, if it runs out, nobody is going to be able to drive over and fit a bigger one. If this causes the system to crash, the results could be catastrophic.

Documentation

Since the Requirements Definition should form part of the understanding, if not the actual contract between the client and developer, it needs to include provisions for nonsystem-related deliverables, such as documentation.

Besides the User Guide, which is destined for the end user as an aid to using the software on a daily basis, there should also be documentation relating to the development and development process. What level of documentation the client wants to receive as deliverables will depend on their technical competence.

Hence, the Requirements Definition should clearly state a cut-off that dictates when a piece of documentation is not required to be signed off on. Both the developer and the customer need to be aware of two things.

First, although one would like the opportunity to sign off on every piece of development documentation, this will invariably hold up the entire process, and cost money. Each time the documents have to go back and forth between the client and developer in the hope of achieving perfection will cost time and money on both sides.

However, liability becomes hard to establish if the client has not signed off on a piece of documentation that later proves that the developer did not understand a key part of the end system.

There is, therefore, a balance to be struck between reading and signing off on every piece of development documentation, and choosing the vital pieces that *need* to be examined carefully to be as certain as possible that the project will be completed successfully.

The Requirements Definition should then allow for:

Design Documents: To show understanding of the basic problem and solution.
Development Documents: To show good software design principles.
Administration Guide: Optional, depending on system design.
User Guide: For the end user.
API Reference: Optional, depending on system design.
Training Guide: For re-educating existing users or training new ones.

with a loose description of what is expected at each level. Not forgetting, of course, that the Requirements Definition and Requirements Specification need to be included in the final list. All of the preceding document areas should be self-explanatory, with the possible exception of the two optional ones.

The Administration Guide only applies to client-server software, or software in which there are different levels of users, each with different responsibilities or permissions. Some examples are operating systems, problem reporting, escalation tools, and project management and tracking utilities.

The other optional item, the API Reference, only applies to those systems that can be extended by further development. In such cases, the future developers who might be extending the system could be different from the original team, or even a different developer altogether. In such cases, it is useful to have a guide to the Application Programming Interface so that it is clear exactly how future development can be done.

Maintenance

There is a tendency to leave the Maintenance arrangements out of the initial Requirements Definition and to rely instead on the developer's standard contractual terms to indicate and enforce liability in the event that changes, either correctional or evolutive, need to be performed at some point after the system has been delivered.

However, maintenance is an important part of the development life cycle if only because a system may be in service for such a long time that errors will be uncovered that were not discovered in the initial release. Moreover, new errors may be introduced as a result of changes made to the system that are also not found until much further down the line.

Therefore, it is wise to include some form of maintenance definition in the Requirements Definition Document. This should include a schedule for determining the cost of both corrective (changes made to fix errors in the initial system) and evolutive (changes made due to a shift in the client's requirements) maintenance.

THE SYSTEM MODEL

We previously offered the following definition of the System model:

"Shows the relationship between system and environment, paying particular attention to where the environment and proposed system need to interface."

In essence, we need to be able to define the requirements of the entire system, in terms of its inputs, outputs, processing, and control functions. After all, to paraphrase the late Douglas Adams, a computer system allows you to interface with the universe and move bits of it around, and so the system model needs to reflect which bits to move, and by how much they are to be moved.

The System Boundary

The logical point at which the system interfaces with its environment is known as the *system boundary*, and deciding which parts of the entire model fall on which side of the boundary is a time-consuming and difficult task. It is also very important to be sure where the line is drawn, since it is difficult to enlarge the scope of the system model once the system design has begun.

On the system boundary, there will also be the opportunity for the system to interact with its environment, which is loosely called the *interface* by system designers. Across this interface, information and actions will flow in both directions—the aforementioned inputs and outputs—and are represented either as data or control paths.

Possibly the most natural and informative representation of the system model is to use a mechanism that [Somerville92] calls the "tabular collection diagram," which represents the system model as a horizontally aligned set of headings, designed to be used to represent the system from various different viewpoints.

We will borrow this theory and couple it with the idea of a system boundary to state that we will represent the System model as a set of interacting ex-system (or environmental) objects. For each object, we will state its interaction with the proposed system via a tabular collection diagram that contains its interface points (input and output), action, and effect.

By way of example, if we are designing a word processor, we might decide that the keyboard is essentially outside of the system boundary—we have no control over it, but need to interface with it to receive input from the user. This gives us one viewpoint of the proposed system, and so we ought to be able to design a tabular collection diagram to contain its definition, as in Table 6.1.

TABLE 6.1   Keyboard Viewpoint Tabular Collection Diagram

Source	Input	Action	Output	Destination
User	Key press	Add letter to document	Modified document	Screen

Of course, Table 6.1 is not exhaustive; we are more interested in the layout than the actual content of the table. Somerville adds another column to the right—Destination—which is the target viewpoint. This may or may not be appropriate, depending on the designer's point of view.

Data Storage

One particular part of any system model is going to relate to data storage—since most computer programs seem to deal with the acquisition and processing or analysis of data, it is logical to spend some time trying to establish where the data is to be stored.

The Data Storage requirements part of the System model need to deal with the following areas:

- What kind of data needs to be stored?
- How much data is to be stored (working data, reference data, and archived data)?

Answering these questions will lead to a number of propositions as to where the data should be stored, and the likelihood is that each type of data, from the set of working data, reference data, and archived data, will be stored in different places, and hence fall in different parts of the system model.

Working data is most likely to be stored as part of the system itself, inside its boundary; reference data may be stored either as part of the system under consideration, or outside as a facility offered by the environment. Data archival is, except in specialized cases, always going to fall outside the system boundary, and be something that has to be dealt with as part of another system. This is more for convenience than anything else.

The volume and type of data will lead to further decisions as to how the data will be stored, both internally and externally. Once it has been ascertained at what point the data moves between the different forms, the decision as to what facility is to be used to store it will probably be reasonably obvious.

Working data, for example, is usually stored in memory, since it needs to be rapidly accessed and will exist in relatively small quantities. At some point, probably after some processing, the data will have changed into a form that is ready to be stored in a more permanent fashion—as a file on the hard drive, for example—at which point it has become reference data. It still needs to be accessed in a timely fashion, but probably only to be consulted, modified, and put back again.

At some point, the data will no longer be current enough to be needed in a timely fashion and can be archived, only to be consulted at irregular, unpredictable intervals. This will usually lead to the data being stored on some kind of removable media, the exact type depending on the volume of information that is to be archived.

REQUIREMENTS AND DEFINITIONS

At first thought, the difference between functional and nonfunctional requirements seems obvious: functional requirements clearly relate to the functions that the system must offer to satisfy the client, and nonfunctional requirements relate to everything else. Of course, it is not always that simple.

Functional Requirements

When we are creating a Requirements Definition Document, we are trying to set out the specific requirements that we are placing on the system from the point of view of the client. Suppose, for example, that we are trying to set out the requirements for a new mode of transport that is to replace the automobile, but offer advantages to the end user over and above that of the existing solution. This is how most consumer software applications begin their lives.

A functional requirement of this new product might be that it is capable of conveying the end user from point A to point B.

Nonfunctional Requirements

All the constraints that define the operating environment that the system needs to be designed to fit into, and around, fall into the category of nonfunctional requirements. Anything that is part of the system can be worked around, using a set of functional requirements. Anything that cannot be altered by adding some functionality to the system is a nonfunctional requirement. This does not preclude a system-oriented solution, but does preclude a functional requirements definition.

For example, if our application software needs to be able to run on a computer system that is resource limited (e.g., it has limited electrical power available) in some way, then this becomes a nonfunctional requirement. There is nothing we can do in terms of functional definitions that will minimize the effect of having to execute on slow hardware.

The eventual software application design will have to get around the problem, but at the requirements level we cannot require that the code is more efficient than it might otherwise be, we can only indicate that the target system is less powerful than one might possibly like.

This is the essential difference between functional and nonfunctional requirements. Functional requirements deal with everything that the system must do, while nonfunctional requirements indicate the circumstances under which the functional requirements will need to be satisfied.

Glossary

Loosely speaking, a glossary is a set of definitions of standard known words, which are specialized, but common knowledge to those dealing with the problem domain on a regular basis. It might also contain some terms that are specific to the system being debated. There are some who feel that these entries might be better housed in the formal requirements definition, rather than in a general-purpose area of the document.

The contents are less important at this point than the way in which the glossary is used. Of course, each entry must adequately convey the required meaning, but it is also important to take into account how the terms are identified in the text of the requirements document, and how the glossary is referenced elsewhere.

The glossary is likely to be worked on by different parties, probably at the same time, because it is a central reference work for the entire project. Therefore, there must be a mechanism by which project members are able to update and refer to it while being sure that their changes are not being lost, nor are they overwriting other people's updates.

This can be achieved by various standard mechanisms, such as ensuring that the document is locked while a project member has it open, or providing for an automated glossary updating system via some form of proprietary software solution.

One final practical note—the glossary will grow as the project progresses. As such, the page number for specific entries will change as new entries are added. This means that if a specific reference exists elsewhere in the documentation, it should be in the form of a "live" link that is updated whenever the glossary changes to ensure that the page numbering remains current. Naturally, this applies to all documents that are centrally managed in this way.

THE SOFTWARE REQUIREMENTS DOCUMENT

A Software Requirements Document is an extension of the Functional Requirements. [Somerville92] quotes [Heninger80] with a list of points that such a document should cover, although they both refer to it as simply the Functional Requirements Document.

(1) It should only specify external system behavior.
(2) It should specify constraints on the implementation.
(3) It should be easy to change.
(4) It should serve as a reference tool for system maintainers.
(5) It should record forethought about the life cycle of the system.
(6) It should characterize acceptable responses to undesired events.

Arguably, we have dealt with many of these points with documents such as the System model (which covers point 1) and nonfunctional requirements (covering point 2), but only at a reasonably abstract, system-oriented level. For our mind, the Software Requirements Document should go further, almost to the point of system specification, in detailing exactly what the system is to do.

Living Reference Document

If we look at the qualitative points from the six noted by Heninger, 3, 4, and 5 are almost entirely to do with the style of the document rather than the actual content that should be present. Point 3 indicates that the document is to be "living" for the duration of the definition process, and hence should be easy to change. This requires a minimum of references to itself that would require updating every time the document changes.

The layout that most facilitates ease of change is to ensure that each thought, diagram, table, or definition can be contained on a single page. If the end of the subject is arrived at before the end of the page, then a new one can be started. In this way, we can be sure that there is plenty of space for additions, and that information will always be easy to find. Of course, it makes for large documents that, if printed, will waste considerable amounts of paper.

However, the counterbalance is the ease of reference, which helps us to satisfy point 4—it is, after all, a reference tool for those who have the unenviable task of trying to maintain the system once it is in production. Unenviable simply because it is very difficult to maintain someone else's system, and good reference documentation is essential in being able to do a good job.

System Behavior

This leaves us with three points relating to the actual system definition itself. Clearly, since this is a document that defines a system in terms of interaction with its environment, only the external behavior of the system is to be noted—as in point 1.

This interaction includes the various ways in which the system is constrained by external influences, again a topic that we have covered in an abstract way, but again something that needs to be elaborated on in the actual Software Requirements Document. For example, the System model might state that there are restrictions on the memory available, but it is up to the Software Requirements Document to stipulate exactly how this will affect the software.

Finally, we need to be able to state what should happen if the constraints are breached in some way; to be able to stipulate the acceptable behavior of the system in cases where it is operating, or attempting to operate, outside its immediate boundary conditions. To continue the previous example—should the system attempt to use more memory than is available, it should recognize this fact and react in a way that is acceptable.

SUMMARY

The Requirements Definition Document, therefore, covers everything that is needed to specify how the system is supposed to react with its environment, in essence, defining its behavior by way of the effect that it has on the operating environment. It also defines everything that will be needed when the Requirements Specifications are written.

The relationship between the Requirements Definition and Requirements Specification is that it is supposed to give a natural language rendition of what the system is supposed to do. The Specification is a more technical document that can be validated. The Requirements Definition needs to be the basis of understanding between the client and developer about what services the system is supposed to deliver.

REFERENCES

[Heninger80] Heninger, K. L. "Specifying software requirements for complex systems. New techniques and their applications." *IEEE Trans. Software Engineering,* 6 (1), 2–13.

[Somerville92] Somerville, Ian. *Software Engineering,* 4th edition, Addison-Wesley, 1992, p. 51–53, 86.

7 Requirements Specification

INTRODUCTION

The previous chapter dealt with a natural language document that serves as an informal agreement of understanding between the two parties as to what the system is supposed to do. However, due to the inadequacy of natural language, and the abstract nature of what it is we are trying to define, it is much more useful to produce a technical document to define exactly what the system must achieve.

A good requirements specification can also be proven correct, and while it may not actually satisfy the requirements of the client, it will at least be consistent within itself. However, unless the client is technically minded (very rarely the case), they will not be able to understand it, making it useless as part of a binding contract.

It is therefore necessary to strike a balance between the over-flexibility of a natural language representation and technical nature of a precise, mathematically formal language. Such a compromise can be found in using a Program Definition Language (PDL), which will allow the specification of the customer's requirements to be precise, correct, and unambiguous.

This chapter leads the reader through the structure and content of the document, which will find its place alongside the Requirements Definition detailed in the previous chapter, as part of the contractual agreement between the client and developer. Once again, we present a skeleton layout that will help the readers to formulate their own version that can be reused and refined as more projects are undertaken.

Many professionals and academics seem to fall into one of two camps—either they group Requirements Definition and Requirements Specification together in one document, or, like Somerville, they separate the two, but then use the terms *Requirements Specification* and *Functional Specification* interchangeably.

The reader will have noticed that we have three separate chapters in this book: Requirements Definition, Requirements Specification, and Functional Specification. The reasoning is that the Requirements documents should be a contractually binding expression of the requirements that the client has in order to produce the system they require.

The Definition document needs to be understandable by both parties, and is a natural language rendition of the requirements, which makes it useable as a binding contract; the Specification document need not necessarily be directly understandable by the client, which, as we explained previously means that it might not be acceptable as a contractually binding document.

The Functional Specification, however, should detail, in a technical way, the exact nature of the functions that will be implemented, in terms of the data that they will need to manipulate, and the inputs and outputs for each function that needs to be implemented.

This is why there are some parts of the Requirements Definition document that we have brought forward into the Requirements Specification since they are technical in nature and therefore should not be included in the Requirements Definition, which is essentially a nontechnical document.

SKELETON REQUIREMENTS SPECIFICATION DOCUMENT

In compiling our list of sections for the Requirements Specification, we have drawn items from two sources. The first is the list of Appendixes that Somerville attached to the Requirements Definition document, and which we have carried through from the last chapter.

The second source of inspiration has been various functional specifications with which the author has been involved. Some of those have contained information that, in the author's mind, would have been more appropriately included as part of a Requirements Specification, in the interests of keeping the Functional Specification purely function oriented.

Functional Requirements Specification: This is a formal elaboration of the Functional Definition of the system; as such, it details the exact nature of the facilities that the system needs to offer to satisfy the real-world requirements of the customer.

Nonfunctional Requirements Specification: A formal description of the boundaries and constraints under which the system will need to perform. While the Requirements Definition might state, for example, that the system needs to provide functionality within reasonable bounds, the Non-Functional Requirements Specification will detail the exact nature of those bounds.

Hardware: A description of either specialist hardware that is needed for the system to perform the functions required by the client in order to serve the purpose that they have identified, or a tight specification of the actual hardware that the system will be required to operate on (the platform).

The latter might more appropriately be placed under the nonfunctional requirements, depending on the exact nature of the system under development.

Database Requirements: Again, details that are probably too technical to be included as part of the Requirements Definition, but will have an impact on the way the Functional Specification is written.

While the Functional Specification will include a Data Dictionary to specify exactly how the data that is required by the system will be manipulated, the Database Requirements will specify what data is needed, and show the logical relationships between those data.

Internetworking and Mass Storage Requirements: These two items might need a little explanation. It is clear that they are of a technical nature, and therefore can be alluded to in the Requirements Definition (using phrases such as "must have e-mail capabilities" or "must support large removable media"), but whose exact technical specifications mean that, following our philosophy on these matters, they belong in the Requirements Specification document. On the other hand however, one might argue that they do not need their own section, and could be integrated within the Nonfunctional Requirements, Functional Requirements, or even the Hardware section of the document.

Since large-scale software development first began, the landscape of computing in general has changed somewhat. First, the advent of the Internet, and explosion of two of its most important components—electronic mail and the World Wide Web—have led to a whole glut of different applications, all taking advantage of what we call here *internetworking*.

The price per megabyte of mass storage devices such as huge (terabyte) capacity hard drives, high-capacity tapes (8 to 18 gigabytes), and regular-sized optical disks such as DVDs (4 gigabyte) and CDs (700 megabyte) means that we have a whole new area of functionality to deal with. This comes with its own special set of concerns, as well as rich features we can exploit.

The floppy diskette (1.4 megabyte), once the only removable storage and backup possibility, fades into insignificance alongside today's mass storage heavyweights.

Since both items rely heavily on external specifications and industry standards, we have chosen to place them in their own section. While the facilities that they offer might form part of the desired behavior of the system, they can be said to be on the boundary of the functionality that the developer is expected to provide, and as such are considered in a separate category to the rest of the system specifications.

The remainder of this part of the chapter will look at each of the previous sections in detail, showing how they relate backwards in the document chain (to the Requirements Definition) and feed into the Functional Specification that provides the developer with a guide as to what the system is actually supposed to do.

Functional Requirements Specification

This section of the Requirements Specification needs to formally define the functionality that the system must provide, but not the actual functions that will support that functionality. By way of example, we should consider a small accounting system, which contains one part allowing for the input of invoices.

The Functional Requirements Definition might state that the system must be able to support the input of invoices, detailing the various pieces of information that must be retained by the system. It might also provide a description of any sorting, retrieval, and reporting mechanisms.

When this becomes part of the Functional Specification, each of the operations that need to support the entry of data into the invoice must be catalogued and unambiguously specified in such a way as to represent the clear movement of data through the system, not to mention a definition of the data that needs to be stored.

The Requirements Specification becomes the bridge between the highly technical and concise Functional Specification and the contractually binding but nonetheless imprecise wording of the Functional Definition.

As can be seen in Table 7.1, each time a specific part of the system is defined—starting with the reasonably vague Definition, through to the contractually binding, but still nontechnical but precise Requirements Specification, and finally to the technical description of what the system must do, the Functional Specification—it becomes expanded from an idea into a concrete set of precise features.

TABLE 7.1   From Definition to Specification

Requirements Definition	Requirements Specification	Functional Specification
Needs to store invoices	Invoices stored alphabetically, by family name, then by first name, where applicable. Company names are stored as family names with status (LLC, etc.) in place of first name.	Invoice storage supported by: —SaveInvoice Purpose … Input … Output … —SortInvoices Purpose … Input … Output … … etc.

One might argue that the steps in Table 7.1 are unrealistic; that is, that the client would not ask for an invoice storage system, and not think to directly instruct the contractor that it should also be capable of sorting them, allow searching, and so on.

It is true that the progression shown in Table 7.1 indicates that the process seems to expand the problem domain dramatically with each step, but it would be unfair to suggest that this is an exaggeration. Quite often, an abstract wish of the client can turn into a substantial part of the project, and as the project team attempts to arrive at an all-encompassing definition, they cover parts that the client has not considered, but will find useful enough to be willing to pay for.

Nonfunctional Requirements Specification

The system will need to be designed to operate within a certain environment, and the Nonfunctional Requirements Specifications need to detail, in precise terms, what that environment will be.

This can include everything that will have an effect on the system, but does not form part of the system itself. If we want to be completely formal about the project specifications, we would need to indicate areas such as the contractual conditions under which the system will be required to operate and any other, possibly more technical considerations.

However, specific pieces of hardware with which the system is designed to interact, and other pieces of software that will provide services to the system under specification can be considered elsewhere—in the Hardware, Database, Internetworking, and Mass Storage sections of the Requirements Specification.

The reader will note that there is a column missing in Table 7.2 with respect to Table 7.1, which showed the progression of the Functional Requirements—which would be the equivalent to the Functional Specification.

TABLE 7.2 From Definition to Specification

Nonfunctional Requirements Definition	Nonfunctional Requirements Specification
Should run on an office standard PC	Platform will be the minimum requirements for Windows XP

In Chapter 8, "Functional Specification," we elaborate on a section called Non-System Functional Specifications that include, among other things, a definition of some of the Nonfunctional Requirements, but only where they offer functions that the system is able or required to use.

Where in Table 7.2 we indicate that the platform requirements will be the same as for a Windows XP workstation, we might then insert in the Nonsystem Functional Specifications the exact specifications of a Windows XP system as used by clients in a similar office environment.

Hardware

In a given system, there will be various types of hardware—ranging from that which the system needs to run on, often called the *target platform*, to any hardware with which the system must interface, and does not have any standard way of doing so.

One might choose to place the internetworking and mass storage requirements in the Hardware Requirements. Both need to use hardware devices to provide the ancillary services for which they are designed. If there is nothing else of note in the Hardware Requirements Specifications, then this approach is quite acceptable.

However, it is always possible that the Hardware Requirements Specification will already be filled with entries relating to the target platform, and specialist hardware interfaces, at which point it becomes cumbersome to mix this information in with other, unrelated entries that one might reasonably expect to be found elsewhere.

Database Requirements

Many systems require some form of database support, and it is quite likely that this will come from a third party, either in the form of a separate component or libraries that can be built into the application to provide the support that is required.

In order to decide which route to take, it is necessary to specify exactly what the features and boundaries of the database system will be. In the Nonsystem Functional Specification, which we will look at in Chapter 8, there will be a certain number of features that will be required of the database system, but it is in the Database Requirements section of the Requirements Specification that we will detail the required capacity of the system, as well as the functions it will need to provide.

Functional Requirements

One of the most important aspects to consider is whether the database system will need to travel with the software under development, or whether it may reside on a separate machine, and be used in a client-server environment. The central storage as opposed to distributed storage debate will hinge on the system requirements.

Central storage is only really required if the data stored has to be constant across all clients connected to the system. For example, a contact list in an office automation package (such as Microsoft® Exchange) should be accessible by all, and be updated by a subset of the users, but always current, no matter who accesses it, and from where.

However, this may only mean that the central storage is encapsulated in a server application (as in the case of Microsoft Exchange), which in turn may use a local database, as opposed to client access to a database server such as Oracle.

This then leads to another potential requirement—a database that uses a language such as Structured Query Language (SQL) to interface with it, or one that can only be accessed via a specific Application Programming Interface (API) such as that provided in Visual Basic toward the Microsoft Access database software.

Using a language such as SQL becomes a requirement if ad-hoc access to the database, programmed by the user, is part of the system definition. This may be necessary in systems that are designed to collect data, in the expectation that the users will extract and report on that data. If an API-based database system is used, then a programmer will need to extend the system to allow the user to query the database in a specific way, should the need arise.

There are also a number of specific functions that might be required of the database system, some of which are reasonably standard, but some of which can be considered advanced functions only offered by the high-performance, enterprise database management systems.

These might include the ability to search the database by using a function of the database system as opposed to retrieving records on a sequential basis, and either retrieving the data or discarding it and moving on to the next inside the software system. Clearly, it is much more convenient to have the database perform the selection—but this might conflict with other requirements.

Another aspect of database systems relates to the fact that they hold the most important business commodity—information. The amount of importance that the client places on the information being stored in the database will depend on what the system is designed to do.

Most modern database systems allow for some form of basic data preservation—such as not actually deleting records until the system undergoes maintenance (such as the compression function offered by the FoxPro™ programmable database system)—and even some protection against system instability.

Another useful function that might be required is the possibility to offer automatic reporting features; not necessarily arranging the data contained within in a specific way, but also reporting on the state of the system, and the interactions between it and the various users of the system.

All of the preceding can be specified in very precise terms, using industry standard terminology to specify the exact requirements that the database system chosen in the implementation needs to satisfy.

Performance

Performance is usually measured using some reasonably standard definitions—similar to those used in specifying the performance of hard disk drives. There tend to be four main criteria for specifying the performance of a database system—capacity, search speed, access time, and update time.

The capacity refers to the number of records of a certain size that can be included in the entire database. While this might be limited in theory by the memory or disk space attached to the system, most database systems are built such that there is a practical maximum beyond which the other three criteria will begin to suffer.

The search speed should be indicated in a meaningful measurement that represents the requirements of the client in an unambiguous fashion. Simply stating that the database should be searchable within a reasonable wait time for the operator tells us that we would prefer not to be kept waiting, and is acceptable for the Requirements Definition.

The Specification, however, requires that more precise terminology is used—possibly something stating the number of results returned per second within a certain time of submitting the query. For systems designed to be used in environments in which response time with complete information is important, such as the retail sector, this is obviously not the correct measurement to use.

In such cases, a more useful way to define the requirement is in terms of returning enough meaningful information within a certain time span. For example, 100 customers, by name and first initial, to be returned within five seconds of issuing the search command. This is also a useful approach when specifying search criteria that could possibly result in many records, but which it will be easy to spot straight away if the set is too large, and the query can be modified.

The last two, access time and update time, are related. The access time, measured by a single value indicates the amount of time that is required to retrieve a single, indexed record (or to insert or delete one). The update time indicates, as a single value, the time taken to change a record as a result of a query.

Nonfunctional Requirements

Anything that does not affect the operation of the database is a nonfunctional requirement. This will inevitably include requirements that relate largely to nontechnical issues, such as price, target platform, and various availability criteria, such as platform, documentation language, and so on.

It is important to include such items as constraints on the purchase of equipment; a client may require a database system that they will be unable to accommodate on their existing hardware systems, but be unwilling to acquire new hardware to provide a platform for the database.

Internetworking and Mass Storage

At a certain moment in time, connecting two or more machines together in a network, or connecting heterogeneous networks together, would have presented enough of a challenge to be part of the product itself. The ability to read and write large quantities of data to CD, DVD, or similar would also have fallen into this category.

Since the advent of consumer-driven technologies such as audio CD and DVD movies, the implementation of these features has become part of the state of the art, in the same way that large, multi-user relational databases have.

For this reason, the Requirements Definition and Specification will not cover these items as part of the discussion of the main product. There will simply be references to the product being "Internet ready" or that it must be able to "back up to CD."

To accurately describe the requirements, we must consider key areas of their functionality that will constrain the end system, in terms of the features that the technology offers.

Multi-User Requirements

Many non-IT professionals tend to think of software as being either single- or multiuser. In other words, it is used either by a single individual on a single workstation, or by multiple individuals on different, possibly interchangeable, workstations.

Increasingly, however, industry is redefining the relationship between the user and the software, which has led to many different layers of single, nonsimultaneous multiuser and fully cooperative multiuser distinctions. The client needs to be aware that, while it is possible to change between these once the development has begun, the design will usually depend on the number of users, and simultaneous users that they expect to be able to use the system.

If, for example, a system has been designed for single users each with a single workstation, and local database, then expecting it to be extended, once created, to include a simultaneous multi-user, centrally managed database is going to be a very difficult proposition.

There are four discrete levels of multiuser capability, based on whether there are multiple users, multiple machines, or both:

- Single user, single machine
- Multiple users, single machine
- Single users, multiple machines
- Multiple users, multiple machines

For each of these levels, we must also consider, in cases where multiple users or machines are foreseen, whether they will do so in a simultaneous or nonsimultaneous fashion. The exact mix will vary depending on the product being designed, as will the consequences.

Those applications that do not require a database or other kind of information store will be impacted less by a move from single to multiple users (or machines), while systems that make use of peripherals, external databases, networks, and other facilities will be impacted in a much more important manner.

Network Protocol

Part of the multiuser, multi-machine debate is solved by deciding on the network protocol that will be used by the end system. There are basically two choices—Internet ready or LAN/WAN oriented. Network engineers might argue that the actual networking landscape is much more complex than this, but at this stage, we can ignore much of the underlying intricacies of network engineering.

For each of the two key network system types, there are multiple possibilities; Internet-ready applications can either be server based, or client/server oriented. They can also be used in a browser, or through a proprietary application serving as an interface to the back office system.

Systems that are networked but not Internet ready can use direct connections supported by the underlying operating system, or they can use techniques such as file sharing to provide the illusion of peer-to-peer interaction.

If the application needs to make use of Internet technology, but not on a peer-to-peer basis, such as allowing for backend connections to Web servers, allowing electronic mail messaging, and so forth, then these requirements need to be laid out in terms of the supporting technology.

Client/server systems are at the top end of the scale—such applications are essentially built up of two or more layers. There is traditionally a server-based system that carries out much of the central processing; such as database access, peripheral sharing, user access management, and so forth.

The user interface is then split over several workstations that form the client side of the system, where each has specific local tasks that it needs to carry out in order to succeed in whatever role it has been assigned.

Security

Security is a big issue for multiuser networked applications, and this is most apparent in those applications that are distributed over large geographical areas using Internet technology. They are open to abuse from both the inside and outside—hackers in the first instance, disgruntled employees in the second.

While the latter is, and will hopefully remain, rare, hacking as a profession is on the increase, and both the client and application designers need to make sure that they have covered all the important security loopholes that might exist in the system design.

Of course, where it will be more difficult to locate and fix security concerns is in the operating system itself, and if this is a key sticking point for the system under development, then the advice is usually to seek the help of a company specializing in the operating system for which the system is being deployed.

Having said that, LAN applications can usually safely use the security mechanisms of the underlying operating system to manage the relationship between the client and the server, or between clients in a peer-to-peer environment.

One of the key security issues is whether to allow users to define their own passwords, use system assigned passwords, or use a hardware token. The first option is the easiest—the system allows the users to choose their own password, but manages the changing of that password and applies some simple rules to determine the suitability of that password.

These can range from simple tests on the length of the password—insisting that it is no shorter than six characters is a common one—right up to stringent tests that make sure that the password consists of letters, numbers, and special characters, and that there are no adjacent letters that could form a word from a list stored by the server.

The second option uses a random, system-specified password that is communicated to the users such that, once changed, they will need to use it the next time they attempt to gain access to the system. Of course, the drawback here is that the user will probably write it down, since it will have been chosen in such a way as to be difficult to guess, and hence, remember.

Finally, the system could use a challenge-response hardware token system, where the user accesses the token with a four-digit PIN and enters a specific chain of digits generated by the server. The token (resembling a pocket calculator) then responds with a string of digits to be returned to the server.

Assuming that the relationship between the two sets of digits is acceptable, the user is permitted access to the system. This option requires the most sophistication, and is the most expensive.

The client needs to carefully evaluate how much security is appropriate, and detail it as part of the Requirements Specification, so that the developer can then decide on the most robust and efficient design, and estimate cost accordingly.

Media Requirements

The purpose of the Media Requirements Specification is twofold: on the one hand, it is needed to try to estimate the amount of temporary storage that will be needed, either at the server, or on each client, but on the other hand, we also need to specify what kind of permanent storage solution is required.

In this case, permanent storage will probably refer to backups of the system, taking the data away from the running environment and storing it safely on more permanent media such as a tape, CD, or DVD.

Which of these is used will depend on:

- Quantity of data to be archived
- Archival frequency and timing
- Existing technology

Different solutions will be required for different environments. A deep space mission, for example, will need to take backups onto rewriteable media, such as magnetic tapes, since burning a CD will waste valuable resources, and the craft will not be able to take enough for the entire mission.

Rewritable DVDs, however, would probably provide a good solution, being robust, compact, and reusable, but their data capacity tends to be lower than magnetic tape. Then again, magnetic tape could be affected by electromagnetic pulses, which abound in deep space.

These are the kinds of client-specific questions that need to be asked, and which the developer, in all probability, will not have the specialist knowledge of the client's problem domain available to him if the questions were to arise during the design phase. Hence, the time to tackle the issues is during the specification phase, when the clients are at their most involved with the project.

PROGRAM DEFINITION LANGUAGE

Rather than relying on natural language to describe the Functional Requirements of a system, some software engineers prefer to use a Program Definition Language, or PDL. It should be mentioned that this might sound as if we are prematurely attempting to address the design and implementation of the end system, but that a PDL is only capable of describing the effects of a system, not how it is to achieve it.

When to Use a PDL

A PDL should only be used under certain circumstances, and such use is constrained by the readership of the Requirements Specification document. While it is true that the PDL defines the system in a way that should be unambiguous to the reader, it is equally true that only clients possessing a certain level of technical understanding will be able to read a PDL and thus confirm its veracity.

A PDL representation of a system is very useful in cases where the system needs to interact with another one that already has a set of rules governing the information and processing that it offers. In such cases, modeling the interface between the two systems using a PDL will ensure that the two systems are compatible at the Requirements stage.

Other systems that will benefit from a PDL include those with a high content of user interaction or logic processing, and real-time systems and mission-critical applications.

What Is a PDL?

Formally speaking, a PDL is a limited collection of constructs and features that can be used to describe the elements of a system in such a way that the logic and data are presented in an unambiguous manner.

These constraints stem from the fact that a PDL is usually based on a real programming language, and is thus provable. Indeed, if the PDL is an actual programming language, it will have tools that can be used to validate the specification built using it.

Choosing a PDL that is appropriate for the system being specified is a difficult proposition. If the final system is to be highly interactive, then a different kind of PDL should be used from a system that has very little interaction, but a great deal of internal logic and processing.

The PDL needs also to comply with any standards that the target organization has put in place for the control of code and documents, and needs to be an effective communication tool with respect to the client.

Typically, at the specification stage, the PDL needs only be effective in managing and validating the functional requirements of the system—describing the externally visible effects and elements of the proposed system, and therefore a limited logic and data system is all that will be required.

The PDL tools will then be able to verify that the requirements have been correctly formed, but it will not be able to indicate whether anything has been left out, only that what is present is consistent.

What a PDL Is Not

A PDL is not designed to replace the actual programming of the system, which needs to follow correct system definition. Merely describing the functional requirements with a PDL will not be sufficient to target the resulting package for the correct environment, taking into account all the nonfunctional requirements as per the software design phase.

Besides the fact that to use the PDL as the implementation of the functional requirements will effectively skip a stage in the software development life cycle, there are other reasons why a PDL description of the system should not replace the design and implementation of the system.

For example, while we noted that the client in receipt of the PDL description of the system should be technically competent, this does not necessarily mean that they will be able to read code written in a traditional programming language.

Thus, the PDL needs to be easier to read than, C code, for example; however, unambiguous enough that systems can be described in such a way that leaves no doubt as to the functionality that they are supposed to offer.

If a real programming language is used, then it should be able to be compiled, or at least tested to ensure that it is without error. In other words, it would be a good approach to choose a close to natural language programming language (such as Ada) that has tools to support it.

Such languages do not usually benefit from the extensive support that languages such as C, C++, Java, and so forth, have built up over the years. This lack of support can mean that the language is not compiled but interpreted, which makes it cumbersome for the development of stand-alone applications, and introduces an additional layer of dependencies to the system—something else that needs to be tested.

Examples of PDLs

While technically, a PDL can be created from any programming language, some variants lend themselves to being adapted in this way more than others do. We mentioned previously that Ada is a useful starting point, as are languages such as Moldula-3, Pascal, and those that tend to use a syntax that is closer to natural language than languages based on C.

The first step in choosing a PDL is to isolate whether the system being specified is inherently object oriented at its base. This has no bearing on whether Object-Oriented design and programming principles will be applied in building the system, but rather whether the real-world system lends itself to being described in an object-oriented fashion.

Ada, Modula-3, and Eiffel are all object-oriented languages that can be used as a starting point for creating a PDL, and all are supported by compilers, interpreters, and other tools that are moderately mature, and hence cost effective and stable.

If, for example, we want to specify that part of the functionality of the system deals with an interface to an address book, we can use an Ada package to create an address book consisting of a series of objects, of which one might be an address record, described thus:

```
type AddressEntry is record
  StreetNumber : Natural := 0;
  StreetName : String := "";
  City : String := "";
  State : USState := "XX";
end record;
```

The preceding example uses a type definition, USState, which illustrates one of the reasons why a PDL is a good specification method, since we can use it to discretely and completely specify a set of data in a way that will mean that it can be validated whenever it appears, against our definition:

```
type USState is ( "AK", "AZ" … "XX" );
```

Note that I have not included all the U.S. States in the preceding example, for the sake of brevity, and that the special value "XX" has been included as a default value. We can construct a similar record in Modula-3:

```
TYPE AddressEntry = RECORD
  StreetNumber : INTEGER;
  StreetName : ARRAY [0..999] OF CHAR;
  City : ARRAY [0..50] OF CHAR;
  State : USState;
END
```

The type definition for USState can be defined using Modula-3 as:

```
TYPE USState = { AK, AZ, … XX };
```

One of the differences between the representations is that, having defined our type in Modula-3, we can then perform assignments as:

```
Some_state = USState.FL;
```

This makes validation easier than with the Ada equivalent, which uses a string representation. It can also be easier to understand for the specification of complex systems.

We can then move on from specifying the data to specifying how that data is to represent the parts of the system identified in the system model.

VALIDATING THE SYSTEM

Once the Requirements Specification has been created, it needs to be validated. All parts of the Specification are subject to validation, no matter how they are represented, or what they are supposed to represent.

Somerville identifies a series of four steps needed when validating requirements, which we can group into:

1. Needs
2. Consistency
3. Completeness
4. Achievability

For each of the two areas of the Specification, we must ensure that the users' needs have been met, that the specifications are consistent with respect to themselves, that they are complete, and that they can actually be realized.

The whole ethos of this book has been geared to ensuring that the client (and by implication, the user) is involved to such a level that the validation should confirm that the needs have been met. The use of a PDL for the Functional Requirements Specification will probably help to ensure that the specification is consistent, but cannot help with validating the completeness.

Finally, the achievability of the specification—in other words, whether it can actually be turned into a design that can then be implemented as an actual system— needs to be addressed. This is an aspect where the developers can find themselves promising something more than the current state of the art, or available budget, can provide, so care must be taken to be objective.

However, it is acceptable to try to predict the evolution of the state of technology in certain, fast-moving technological environments.

Functional Requirements

The one aspect of the Functional Requirements that ought to be easy to validate is the consistency. Luckily, in many cases, and in most where a PDL is used, this should be the case. After all, part of the reason for using a PDL is so the system model can be verified against itself.

However, addressing user needs (offering all the functions that they require), checking for completeness (for all the defined functions, are they completely specified), and ensuring that the system can be achieved are less straightforward, even for functional requirements.

Checking that the proposed system offers all the functionality that the user requires, rendering it useful and less prone to being put to one side after delivery as being incomplete, is a manual task. The client must read the specifications, understand them, and match them against the real-world problem they are trying to solve.

This means that the users must be involved, and therefore the document has to be watered down for presentation to them, since in most cases it will be too technical in its available state. This also introduces a layer of uncertainty, since the Specifications will have been interpreted by the client before presentation to the user, and therefore questions or comments that the users might have could be relating to non-existent problems raised by a misinterpretation of the Specification document.

However, it is a necessary part of the validation phase, since we must be certain that we have captured all the requirements during the Definition phase (described in Chapter 6, "Requirements Definition"), and accurately specified them in an unambiguous manner. It will be a time-consuming and relatively expensive process, but one that will save unnecessary intervention later in the development process.

Once we are sure that the Requirements represent the users' needs, we can then look at the consistency of the statements contained within the document. If a PDL has not been used, then this is another manual task that has to be performed to check that there are no conflicting statements or definitions.

If the Specification is consistent, we can proceed to validate the completeness with reference to the user needs. This validation should be performed by the system designers, who should try to ensure that, taking into account the user needs, the requirements of the system have been properly addressed, including those that have not actually been voiced by the client but are a consequence of another requirement.

Finally, care must be taken to make sure that the functions are achievable with the current state of the art, or whether new advances in technology will be required to support the functions that have been isolated. These advances should be anticipated, or the scope of the project reduced to avoid failing this validation.

Nonfunctional Requirements

Testing nonfunctional requirements against user needs is reasonably straightforward—especially where system constraints are concerned. After all, if the system needs to fit into 512 kilobytes of memory, because that is the amount available on the target device, then it is easy to express and validate. If at any time the memory requirements are exceeded, as a result of adding together all the space required by the data types and representation laid out in the Functional Requirements, then this is a clear indication that the user needs have not been met.

Furthermore, it may transpire that the required functionality cannot be delivered in such a manner as to fit into 512 kilobytes of memory, as described by the system. At this point, there are two options: redevelop the user platform, or alter the scope of the system.

It is far cheaper to find this kind of constraint violation during the Requirements Specification stage than at any other stage during the project life cycle. As this simple example shows, validating the nonfunctional requirements can also be an effective cross-check against the functional requirements and a way of ensuring that the user needs have been addressed properly.

Again, though, only dialog with the client and users will enable the developer to be sure that the description of the nonfunctional requirements is complete. There may be pieces missing that are vital to the achievability of the system, but are overlooked during the first iteration of the Requirements Specification document.

Care also needs to be taken to ensure that the nonfunctional requirements can be validated; that is, that the language used to express them provides benchmarks against which the result can be measured. In a sense, they then become as testable as the actual program code.

For example, rather than discussing aspects of the system in vague terms such as "easy to use" and "efficiently," "effective," "high capacity" and so forth, exact numerical thresholds below which it would not be allowable to exceed must be placed on the requirements.

Technical nonfunctional requirements, such as those relating to performance and capacity, are easy, as we saw previously, to quantify. Human or user requirements are less easy to address. If we are discussing a user interface for data entry, we might want to express the fact that:

"Minimal training should be required to effectively input information."

This cannot be tested because it puts no limits on what might constitute minimal training, or what is considered effective. Putting to one side the fact that the effectiveness might depend on the user, we might rephrase the preceding statement as:

"After 15 minutes of formal introduction to each data entry screen, a user of average competence should be able to input 10 records per hour, with a 99.8% accuracy rate."

The reader will note that in the preceding statement, we introduced another vague concept (average competence), which serves to illustrate the fact that it will never be entirely possible to encapsulate the nonfunctional requirements. Since they serve as a part of the contract in many cases, it might be wise to try to narrow down the definition of competence so that it can be measured.

SUMMARY

While many texts on the subject put Requirements Specification and Functional Specification under the same heading, we have chosen to split them into two chapters in this book since there are aspects that have different audiences.

Part of the problem is that the Requirements Specification needs to be able to be understood and interpreted by a nontechnical audience, the client, and will probably form part of the contract. As such, it might be a better approach to place the Functional Specifications in a separate document that is aimed at the designers of the system, who should be technically competent to read a highly abstract document.

We also covered interfacing to other systems, whether they are systems that need to offer services to the end system (such as databases or web servers) or existing systems with which it will be necessary to dialog in order to provide the functionality that the client requires.

We also covered PDLs as an effective mechanism for specifying requirements, and while this approach encroaches on a Functional Specification, the next chapter should serve to indicate where the two documents differ.

Finally, the importance of validating the specification was discussed, which is a concept that is also applicable to many other aspects of software engineering. As long as we validate each phase of the cycle, we can be sure that the result will be of an acceptable level of quality.

8 Functional Specification

In This Chapter

- Introduction
- Process Descriptions
- Data Dictionary
- Nonsystem Functional Specifications
- From Requirements to Specification
- Summary

INTRODUCTION

While the Requirements documentation is clearly aimed at establishing an understanding as to the purpose of the system between client and developer, the Functional Specification is aimed squarely at the development team. It should be a reduction of the Requirements documentation to technical terms, encompassing enough detail as to leave no doubt as to what purpose the system is supposed to serve, such that the end product matches the client's expectations.

In essence, the Functional Specification boils down to defining the data that will represent various areas of the system, and the operations that need to take place on that data for it to result in the system that satisfies the requirements laid out as the contract between the client and developer.

It should be a balance between data models, process diagrams, and natural language annotations, but is not a system design; there should still be nothing that indicates how the data becomes transformed by the operations laid out in the document, merely that a transformation takes place.

The Functional Specification also provides a description of the system that can be used by a third party should they need to reuse parts of the system, or communicate with the system from the outside, without having to dig through the source code or design documents.

Since the aim of the Functional Specification is to define what the system does, and not how it is to be done, it is an ideal document for sharing with other developers who might want to take advantage of the features that it offers, or as a starting point for the development of other systems to perform the same task, but with different implementations.

This last is especially applicable to hard real-time systems in which there is much at stake, such as those systems controlling aircraft, or managing the workings of a stock exchange. When these are developed, it may be appropriate (as in the case of the U.S. Space Shuttle System) to develop the software several different times, to the same specification, but using different design and development teams.

The three pieces of software are then run in parallel, and each decision is monitored by a fourth system that is charged with deciding on what action to take in the event the result of the three decisions is different depending on which system has produced it. Without a sound Functional Specification, this approach would not be possible, and so it has become an important part of any software development paradigm.

PROCESS DESCRIPTIONS

The Functional Specification needs to offer a description of what functions the system is to provide, and how these functions interact to provide processes that operate on the various pieces of data that exist in the system.

We have already dealt with how we arrive at a suitable definition for the system, via the Requirements Definition and Requirements Specification, but these did not narrow down the concepts to concrete functions that can be designed and implemented.

Process Diagrams

One of the best ways to describe how processes interact is by using process diagrams. These are distinct from the data flow diagrams that we have already met, as they give a step-by-step specification of what a single process needs to achieve. They may, however, need to show the movement of data, but this is not their primary function.

Flow diagrams are among the best kind of descriptive diagrams that can be used in the visual description of a process. They should remain easy to follow, absorb, and validate, with the help of use cases. Therefore, before we go any further, we should indicate what each Process Diagram entry should contain:

- The Process Diagram
- A reference to Data Entities and Processes used
- Use Cases
- Test and Validation results

For example, in an accounting package, we might have a process that is needed to calculate the amount of tax to be charged on a specific item. This tax will change depending on what kind of product is being sold, and so we need to perform a few steps, as shown in Figure 8.1.

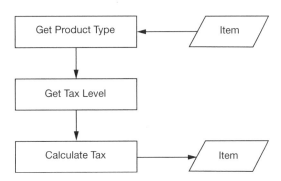

FIGURE 8.1 Tax calculation process diagram.

In addition to the process diagram, we will need to refer to the Data Entities and Process Descriptions so that the reader can cross-check that there is indeed a data structure that holds the required information, and a process that can look it up for us.

Then, we need to present some Use Cases that will show how we expect the process to be used, and how it should react in various possible situations. These can be used to validate the process, using test data that should be constructed specially for the purpose, and used in hand execution of the diagram that accompanies it.

This is distinct from the validation process, which tests the identified boundary cases to check that the behavior is consistent with the requirements of the customer, and that the failure of the process is handled correctly. Boundary cases, as we will discuss later, refer to test data that tests cases that are valid, but only just, as well as those cases that are on the invalid side of the boundary that represents acceptable use.

Function Definitions

The Function Definitions part of the Functional Specifications is a list of functions that will need to be developed in order for the software to deliver its intended result. The Function Definitions are both a link between the Process Diagrams and the data dictionary, and the link between the Requirements and the Software Design.

Thus, the Function Definitions need to show:

- The name of the function
- The process to which it belongs
- The operation that it will perform
- The data on which it will operate

The Function Definitions are as close to the software design as it is possible to get, without actually attempting to prescribe the way that the software will be implemented; they still need to concentrate on the "what," and not the "how."

However, some specific software applications will require that specific calculations be performed, or some very detailed process be spelled out step by step, and this can be seen as encroaching on the "how" of what the system is supposed to achieve. In such cases, it is permissible, but has to be justifiable. It is not acceptable for those involved in creating the Functional Specification to try to dictate technology or implementation issues without a direct requirement link given by the client.

DATA DICTIONARY

A Data Dictionary contains specifications for all the data that needs to be passed around the system, and is a reference document that reduces the data models and associated textual descriptions to a list of data objects and subtypes that need to be represented in the system design.

The Dictionary should consist of several sections:

■ Conventions
■ References
■ Data Entity Descriptions

The Conventions section of the document should outline any specific conventions that are used in the document, such as acronyms used for data type expression, such as "an" used to represent data that can be alphanumeric. Other conventions might include those relating to compound fields, such as variable-length fields with a length indicator, and nonstandard data types.

In addition, system specific conventions such as character set representation (ASCII, ANSI, EBCDIC, etc.) and any other items that form part of the nonfunctional specifications but have an effect on the definition of the data types that will be used to represent information in the system.

The Conventions part of the document should be held somewhere centrally, in template format, so that it can either be reused directly in the Dictionary under construction, or referred to in the References part of the document. This References section needs to specify where the readers can find additional information that will help them in understanding the Data Entity Descriptions, which will make up a large part of the final Dictionary.

Data Entity Description Format

As with all documentation, it is important to establish a standard format and apply it to each project. The reasons for this are that, if the specifications for the object, system, or part of the system are to be reused, it will be much easier for other developers to decide whether it fulfills their needs, as compared with other reusable artifacts if all the documentation is standardized.

Key Elements

As a minimum, each data entity needs to be defined in terms of:

■ Name
■ Data length and type
■ Description of use
■ Allowed values
■ References

The name should be the agreed system name for the data entity, and may or may not bear resemblance to the eventual data type name that is used when the specified entity is designed and implemented. It does not, therefore, need to adhere to any specific programming language conventions regarding allowed characters (such as spaces and numbers), and can be as long and as descriptive as required.

As with the name, the data type description can be programming language independent, and should indicate the nature of the data that is designed to be stored in the data entity, using any acronyms and specific data types that have been introduced in the conventions section of the Dictionary document.

Once the name and type of data has been specified, a description of what the entity is to be used for should follow. It should be a plain language description that conveys the circumstances of use and nature of the data, and any restrictions on what information can be placed in the entity.

This last is distinct from the type of data—it should be obvious that alphanumeric data is not appropriate for storage in an entity that has a data type of either alpha or numeric only—the restrictions that we refer to here reflect values that have been reserved or are simply not logical; a subformat for the data (such as a postal code format) is one example.

The last part of the strict data definition is a list of allowed values, if appropriate, and the circumstances under which they may be used, as well as a default value for cases where a function has rendered the contents of the entity as either undefined or unusable. This can happen in data entry software where the user has neglected to offer a required value, and the system chooses to insert something in the entity that can be recognized as nonstandard data.

The allowed values also give clues, beyond the data type itself, for the testing team, as to what kind of tests they should carry out in terms of valid, invalid, and out-of-format data that can be thrown at the system in an attempt to force it to display undesirable behavior, or checks that need to be done to be sure that the system has processed the data properly, and according to the Process Description.

Finally, each entry should contain a list of references that will help the reader to connect the data entity with other parts of the Data Model or Dictionary. If the entity is a small part of a larger object, such as the street name in an address entity, then this should be indicated in the references.

In some cases, it might make more sense, when dealing with many multipart data entities, to take one of several approaches. The first is to divide the Dictionary entries into those that consist of compound entities, and a second that contains the primitives that are used by the compound entities.

This introduces an implied hierarchy, which might begin with very simple data types (characters, numbers, etc.), builds up to more complex ones such as strings, and then finally to the useable system data entities, such as names, addresses, vehicle types, and so forth.

However, one might take the approach that the primitives (strings, characters, etc.) be defined in the Conventions part of the Dictionary, and then the entire entity described in the Dictionary proper.

There are also a number of possibilities in between, and whichever one is chosen will depend on the nature of the system, part of system, or object under specification. Assuming that one is attempting, following the paradigms laid out in this book, to establish some kind of corporate repository of specifications, designs, and implemented objects, then different objects will probably require slightly different specification formats, especially in the area of the Data Dictionary.

Example

The following is an example taken from the definition of data types used in an accounting package:

> Name: Product Name
>
> Type: as25
>
> Description:
>
>> Contains the name of a valid product, as defined by the user. Products can be added at the user's discretion.
>
> Allowed Values:
>
>> Any alphabetical character, or the special character "space".
>
> References:
>
>> See Process Diagrams 1.5, 2.5, and 5.3

We might then go on to define a product entry:

> Name: Product Entry
>
> Type: Record (Product Name, Product Type, Price, Stock, Tax Rate)
>
> Description:
>
>> Contains all the details relating to a specific product, as defined by the user. Records are indexed on a combination of their Product Name and Product Type.
>
> Allowed Values:
>
>> Any combination of values allowed in the specific subtypes. Product Name and Product Type may not be empty. Price may only be zero in specific cases.

References:

> See Data Entries Product Name, Product Type, Price, Stock, Tax Rate
>
> See Process Diagrams 1.6, 2.6, and 5.4

An alternative specification might be to use a predefined compound type to represent the same entity (Product Entry) but without the need to define subtypes such as the Name, Type, Price, Stock, and Tax Rate. This alternative approach might yield an entry like:

Name: Product Entry

Type: TLV

Description:

> This entity contains TLV entries representing product information:

Tag	Length	Value
N	25	Name of Product (as25)
T	25	Product Type (as25)
P	8	Price (n8)
S	12	Available Stock (n12)
X	4	Tax Rate (n4)

> Data may be populated by the user via an appropriate interface (see References) and records are indexed on N25 and T25.

Allowed Values:

> N25   Any alpha character, plus space, may not be empty
>
> T25   Any alpha character, plus space, may not be empty
>
> P8   Signed value, implied decimal point at two places ($105 = 1.05$)
>
> S12   Unsigned value, no decimal point
>
> X4   Unsigned value, implied decimal point at two places ($1250 = 12.50$)

This alternative representation may be found by some to be less easy to read, but somehow more detailed and formal than the previous examples. It is largely up to the corporate style of the organization planning to use this technique as to which specific form to follow.

Notation Standards

There are many different industry standards for data type notation, largely the result of work done by the International Standards Office (ISO) and the American National Standards Institute (ANSI). The former is generally responsible for industry as a whole, while ANSI standards tend to be more or less restricted to computer communications.

Nonetheless, different industries will also have found different ways to standardize the information that they want to have represented, and the ways in which that information is gathered, processed, and stored—such as the Motion Pictures Expert Group (responsible for formats such as MPEG and MP3), the Joint Photographic Expert Group (JPEG image format), and individual companies such as CompuServe (GIF images) and Adobe (Portable Document Format, PDF).

Since, in the computer industry as a whole, we are largely concerned with the translation of analog data into digital data, it makes sense to have a raft of standards for dealing with the more complex forms of compressed and uncompressed data.

For internal use, however, there is a need to standardize notation within an organization so that each engineer shares the same representation for major primitive data types. This has very little to do with the way in which the data entities might be implemented or stored, but has to do with the way in which the type of data that they need to contain is specified. We concentrate on the "what" and not the "how," except in certain circumstances.

One of those circumstances is in the sharing of data, or the transmission of data from one system, or area of a system to another, where one cannot control the implementation of the system with which one is communicating. In such cases, we need to describe an external format that needs to adhere to shared specifications.

Example

In the financial industry, there is a standard for the exchange of messages that provide information relating to financial transactions between two institutions. Each of the major institutions uses the specifications, provided by ISO, as a base from which they can build their own messaging systems.

The ISO standards define each message as consisting of a series of fields, and each field has a specific format. It is largely up to the institution using the ISO Specifications as to how strictly they adhere to these definitions. By and large, however, there are a number of conventions that they do respect.

One of these is the way in which variable-length fields are dealt with; using a length indicator, which can be encoded in a variety of ways, to communicate the length of the data following. We might choose to implement a data representation that looks something like:

```
LL + an
```

This tells us that the field is variable length, has a two-byte length indicator, followed by alphanumeric data. We would, of course, need to define LL and an in the Conventions part of the document. The length indicator can be encoded in a number of different ways:

Numeric Bytes	00–99	100 items
Hexadecimal ASCII Coded	00–FF	255 items
Hexadecimal Bytes	00:00–FF:FF	65,535 items

Since this is rather ambiguous, it is a good idea to choose two of the preceding as standard ways of encoding data, and give them different notational standards:

LL	Numeric Bytes	0–99	100 items
L2	Hexadecimal Bytes	00:00–FF:FF	65,535 items

It will not usually make sense to use the second variation, since this can be represented as:

L1	Hexadecimal Byte	00–FF	255 items

These come into their own when we begin to build more complex fields, such as those containing variable amounts of variable-length data. ISO has chosen to call these TLV fields, which stands for Tag Length Value.

The standard itself leaves the actual implementation wide open, so that organizations can specify their own meaning for the three components. Thus, we might decide that a TLV item be defined as:

```
an2 + L1 + <value>
```

Following our notation, this means that each item has an alphanumeric tag, 2-bytes wide, followed by a hexadecimal length indicator allowing a length of up to 256 bytes, with an unspecified value as the final component.

This is then used to specify individual tags, such as:

```
CC 02 A
```

This means that the tag "CC" allows data of two bytes, which must be alphabetical, capital letters only. It is a possible definition for an ISO country code tag:

US	USA	GB	Great Britain
AU	Australia	SA	South Africa
etc...			

One final point to note is that while the TLV field is variable length, individual tags cannot be—they are fixed in length, and therefore some standards relating to padding out shorter values need to be implemented.

This example was included to show the various pitfalls that may be encountered when one is attempting to set down notational standards, and is not an endorsement of the solutions presented for dealing with communication of variable-length data between two parties.

NONSYSTEM FUNCTIONAL SPECIFICATIONS

When the Functional Specification is mentioned, people invariably think of the technical aspects of the system. This is no different when we consider Nonsystem Functional Specifications; automatically, people start imagining items such as the systems that it will have to interface with, the hardware it will need to run on, and so forth.

These are undeniably an important part, as are measurements relating to processing power, performance, and capacity of the system, but equally important are the nontechnical aspects such as operational concerns, contractual obligations, and documentation specifications.

When we define the Functional Specifications, we are often talking about a system, and not simply a piece of software that may very well only be a part of the overall system being put into place. The NASA Space Shuttle System has a set of specifications that cover everything from the way the heat resistant tiles are constructed to the garments that the astronauts wear and the software controlling the craft.

While it is clear that this book is aimed at software construction, we need to always be aware that the software has to fit into a larger system. We saw this in Chapter 6, "Requirements Definition," when we tried to break down the system for the Requirements Definition.

The heat resistance of space shuttle craft tiles may not be of direct importance to the software controlling the craft during flight, but the fact that there is an upper limit to the temperature that they can withstand, and that the flight control software might be able to adjust the flight path such that it is not reached, is important, and has an effect on the way the system should react, and therefore needs to appear in the nonsystem functional specifications.

Technical Specifications

In Table 7.2, we saw how the Nonfunctional Requirements Definition translated into a Nonfunctional Requirements Specification, and we mentioned that the Functional Specification should take this one step further, by elaborating on the Requirements in specifying exactly what functionality, external to the system, needed to be implemented to support the system. This is what the Nonsystem Functional Technical Specifications are.

Different systems will require different topics to be addressed, and the reader will likely find topics that we have not covered here which they would need to address. There is also a fine line between this part of the Functional Specifications and the Nonfunctional Specifications defined in the previous chapter.

To reiterate, this part of the Functional Specifications need only refer to those functions that are offered by systems outside of the system under development, but which it needs to make use of in order to operate correctly. The classic example is the target platform; it consists of a nonfunctional part, and a functional one—there may be functions offered by the platform that the system needs to use.

System Capacity and Performance

While this might be covered in the Nonfunctional Requirements Specification, giving guidelines on minimal capacity overall, there may also be specific functions that need specific system capacities and performance to function correctly.

We could state, for example, in the Requirements Definition, that the system should be able to be run on a standard office PC, and in the Requirements Specification, agree that this be equivalent to a machine capable of running a specific operating system.

In the Functional Specifications, we might find that we need to narrow this down further to state that a specific function should execute in such a manner that it is completed within a specific time frame with reference to the power of the underlying system.

External Systems Communication

There will be times when the system needs to communicate to other systems, be they databases, networks, printers, and other peripherals, or something entirely different, and the precise way in which such communications should be carried out needs to be defined in the Functional Specifications.

The Data Dictionary will probably contain references to data entities that can be used in such cases to either represent the result of such an interaction, or the data that needs to be passed to the external system to initiate the interaction.

In the Nonsystem Functional Specifications, these definitions need to be expanded into a functional overview of the way in which the interaction will take place. It should look at the external systems as black boxes that fall into one of three categories:

- Accept information from the system (input only)
- Transform information from the system (input and output)
- Produce data for use by the system (output only)

The external systems that are specified will look upon the system under specification in the same way—as an input, input/output, or output entity—usually following a relationship that complements the behavior of each system. In other words, a system producing data and passing it to another system that has defined a relationship as being "input only" is unlikely to return transformed data to the external system.

These relationships, along with the type of data that is to be shared, make up an important part of the Nonsystem Functional Specifications. Without them, the system will not be able to communicate with any other external systems. This might be appropriate, but in most cases, it will not be.

One final note is that this section of the document specifies features for use by other parts of the system under specification. In other words, there should be nothing in the document that is not required by another part of the system. A network interface specification, for example, is not required if the system has a network connection, but never uses it to connect to another system.

However, there may be a function specified that implicates a peripheral such as a printer that will have a specific communication interface that will need to be specified in the nonfunctional specifications.

Nontechnical Specifications

Everything that does not implicate the direct use of IT equipment, but offers a service or function of use to the system or to those using it, is a nontechnical requirement, and needs a nontechnical specification.

This covers a wide-ranging collection of topics, which we have chosen to separate into three key areas:

- Operational Concerns
- Contractual Obligations
- Documentation

Since this book tries to offer as many guidelines for as many different kinds of systems as possible, it is inevitable that we also touch on specifications that are designed for use by third-party developers who want to implement the system themselves, but be able to interact with other implementations of the same system.

This is also implied by the way in which the software engineering problem has been approached in this book; reuse and rapid application development from objects glued together with logic requires that each entity that is developed is, in a sense, initially over-specified.

Indeed, it will need to be, in the first instance, because while it may remain a small and inconsequential part of the system, in the future it may be extended to cover more areas of functionality. Hence, the formal corporation-wide methodology has to be used even for simple objects, which will lead, in some cases, to superfluous specifications being written.

It is better to be in this position, however, than the other, in which the system has not been sufficiently well specified, and therefore it becomes harder and more expensive to locate and fix errors, and renders reuse of the objects that make up the system all but impossible.

Operational Concerns

Around the system, there will be processes that involve the running of the system in a nontechnical way. We have mentioned that systems do not usually exist in a vacuum. There will be other systems with which they need to interact, and these systems might be technical in nature (other computers, networks, etc.) or nontechnical.

Some of the nontechnical operational concerns will be dictated by the environment in which the system is running. There will probably be a number of rules and regulations that govern the systems under the jurisdiction of the system operator, such as access timetables, scheduled backups, installation and change management procedures, and so forth.

The remainder of the nontechnical operational concerns will relate to the way in which the client wants the system be operated, either with reference to any functions that need to support the operation of the system, or with reference to the environment in which it is designed to run.

This last is particularly important in cases where the client is purchasing the system from a developer, but also outsourcing the operation of the system to a third party who is related to the other two by contract only. Thus, in such cases, the Functional Specifications serve to indicate what is expected of the application operator and the developer.

There are many situations in which this might be of interest, but by far the most common will be in systems that are created for the Internet. It is often the case that the systems are specified by the developer and the client, and then handed over to an Internet service provider (ISP), be it a World Wide Web host or some kind of shared server park.

In such cases, the host may have a number of systems to look after, each of which will run either on a dedicated system or alongside other systems. To be able to offer the best service possible, the host will need to know what is expected of the system, and of their own systems, and the Functional Specification is where this information will be listed.

Contractual Obligations

Contractual Obligations need only be listed if they have a direct effect on the functionality of the system. Those pieces of the contract relating to the payment for work done by the developer, including the Functional Specifications itself, warranty and maintenance thereof, are not necessarily going to be included in the Functional Specification.

However, if the system being specified is an attempt to provide a global umbrella for the operation of a collection of, separately developed, software applications, such as a collection of intercommunicating financial institutions, then there will be certain obligations that they will have toward each other, and these will need to be listed in the Functional Specifications.

Such obligations will have an effect on the smooth running of the system as a whole, since if one party fails to fulfill them, the entire operation might become jeopardized.

There might also be contracts that restrict the use of certain components or third-party services that need to be taken into account because they have an impact on the functionality offered by those external components. In such cases, it must be clearly stated, within the Specifications, what those restrictions might be.

While these items might not be immediately relevant, they need to be borne in mind for the overall duration of the project, which will include future modifications, extensions, or cases in which the contractual obligations toward third parties might become relevant to the functionality being offered.

Documentation

Finally, all the documentation that the developer is supposed to deliver, under the terms and conditions of the contract agreed before the contract was awarded to the developer, needs to be listed and specified in as much detail as possible. Here again, though, it is only required to specify those pieces of documentation that have a direct consequence for the functionality of the system. For example:

- Installation of the System and Third-Party Components
- Programmer Interface

Pieces such as the User Guide are probably not relevant for detailed specification in this document. Besides simply listing the documents to be delivered as part of the system, the client and developer should also agree to a Specification that covers the way in which the information is to be presented.

This can include a specific format, and be as detailed as appropriate. It will probably not be necessary to specify details such as the font and font size to be used, unless the client is a governmental organization that has its own set of standards that must be followed.

What can be quite important is the format in which the documentation should be delivered—such as a plain-text file, a word processed document (and if so, which word processor) or a portable, but closed, format such as PDF from Adobe, which can only be edited with specialist tools, and can be locked by the developer.

FROM REQUIREMENTS TO SPECIFICATION

In terms of the functionality of the system, the Functional Specification can be seen as the end of a process that takes several stages and iterations to complete, but delivers a document that is as close to the actual design and implementation as it is possible to get before plunging into the development process itself.

The process is governed by measured progress through the following phases, each of which has success measured by the generation of a suitable document, as described in previous chapters of this book.

Requirements Analysis: This process needs to establish exactly what the client needs the system to be able to do, and within what infrastructure and environment it has to operate.

Requirements Definition: Having analyzed the problem, and appropriate prerequisites, the Requirements Definition puts this information into very clear language that includes the precise limits that define the boundaries of the system.

Requirements Specification: The final part of the Requirements Engineering process is to create a document that formalizes and quantifies the requirements. Note that these three have domains that overlap, and in some cases, only a Requirements Specification will be needed, depending on the exact nature of the system.

Functional Specification: The Functional Specification takes the Requirements Specification and breaks down the problem domain into chunks that can be logically grouped together to provide the expected functionality of the system, by description of their interfaces, both to each other and the outside world.

Design: At the design stage, the "what" of the system is being translated into "how"; we know, in intimate detail, the functions that the system has to fulfill, but it is only at the design stage that we begin to try to define how it is to perform those functions.

Once the Design has been created, the product can be implemented, tested, and delivered; which represents the culmination of the entire software engineering process.

SUMMARY

By now, the reader should have a clearer idea as to why we have chosen to present the Functional Specification as a separate document rather than merely grouping it together, or using the term interchangeably with the Requirements Specification.

There is nothing to prevent the reader from making the decision to produce a single Requirements Specification document, with the Functional Specification as a section within it, although it might become slightly unwieldy. In addition, there might be sections that have been placed in the Functional Specification that the reader would rather include in other documents.

It is hoped that the format presented here is flexible enough to allow the readers a certain amount of freedom of choice regarding which sections belong in which document that they intend to provide as part of a contractual understanding.

There is also a sense that to include the Functional Specification as part of the documents that are bound under contract, and thus form the basis of an understanding that can be broken before the contract is signed, leaves the developer at a disadvantage.

In other words, the development of the framework governing the agreement represents a large amount of work for which there may not be any compensation, and so the developer might prefer not to carry it out until the contract has been signed and part of the payment made, in part to finance the creation of an agreeable Functional Specification.

It is, nonetheless a very important piece of work, which provides the bridge between the client and the developer. This bridge needs to cover everything that the system is supposed to do in such a way that the design that is created from it represents how the system is supposed to provide those functions.

9 The Object-Oriented Paradigm

In This Chapter

- Introduction
- Choosing a Paradigm
- Object-Oriented Design
- Object-Oriented Programming
- Languages
- Object Testing
- Summary

INTRODUCTION

So far, we have discussed an organizational framework for software development, and a robust system by which we can be sure that the client's actual requirements can be translated into a set of documents that enable them to be correctly communicated to the designers. What we have not touched upon is how the system is supposed to satisfy those requirements.

There are several ways to approach the question of designing a system to satisfy a set of requirements, each with its own merits. We have chosen to concentrate on the Object-Oriented (OO) paradigm for the purpose of this book, with reference to other paradigms where appropriate.

The reason we have picked the object-oriented analysis and design process is that it has become one of the best supported in software engineering circles, by virtue of the fact that it reflects the way in which we naturally think about the real world that we are trying to model.

CHOOSING A PARADIGM

When trying to choose a paradigm with which to design and implement the software, it is important to realize that this will naturally be a decision process that needs to take into account many different factors:

- Expertise of development staff
- Problem domain
- Legacy support
- The programming language tools currently in place

These are mentioned in order of relative importance. It is more important to pick a paradigm that the development team will be comfortable with than one that most accurately fits the problem domain. Likewise, support for legacy toolkits, libraries, or hardware needs to be considered before attempting to rewrite the entire system in a cutting-edge language that requires a completely new set of tools, or attempting to make do with existing tools, when the problem domain is crying out for a specific paradigm to be used.

We also need to be aware of what the various paradigms are. We assume that the reader is at least partially familiar with the various software development life cycle processes, and that this book has tried to present a process that is based on the classic "Waterfall model," but with elements of prototyping and object reuse built into it.

With such a process, which tries to encompass the needs for documentation, progress, quality, and reliability, choosing a development paradigm to fit is always going to be difficult. There are essentially three ways to approach the problem:

- Design by Data
- Design by Process
- Modeling

That is, we can choose to create a system in which the data is accurately modeled, and build it around the manipulation of that data, or we can design the system around the processes that need to take place to satisfy the problem presented by the client. A final option is a combination, in which the system is viewed as a model of a real-world process, containing objects that interact with each other—it is this option that we have chosen to elaborate in this chapter.

However, since the OO paradigm is the natural progression from Data and Process based design, we will first look at these predecessors, as they both share techniques that are useful in performing Object-Oriented analysis and design.

Data Structure Diagrams

If one is going to use a paradigm that is data centric, then some form of diagramming will be required almost immediately. It is far easier to *draw* a description of the data than it is to write about it. This is because the human mind needs to be able to abstract the real-world representation and come up with something that represents the data without actually being the data.

In Figure 9.1, for example, we have a diagram that shows that a piece of data is made up of three subpieces. It could be an Address Record, made up of a name, address, and telephone number, with each Address Record having exactly one Name, Address, and Telephone part.

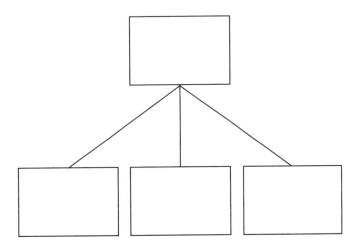

FIGURE 9.1 Data structure diagram.

Of course, each of these parts is also going to have subpieces, and at some point, we need to be able to express the possibility that a number of such subpieces are associated with a particular part of the data system. Therefore, the notation has been augmented with a number of symbols that are designed to indicate cardinality in the relationships.

Figure 9.2 shows the various flavors of cardinality that have been introduced. With these, we can specify, for example, that a car has four doors, without the need for four boxes hanging off the car, each representing a door. Of course, we will probably want three boxes—front doors, rear doors, and trunk—but the idea of cardinality remains.

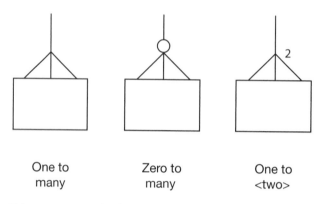

One to
many

Zero to
many

One to
<two>

FIGURE 9.2   Cardinality notation.

Breaking down the information that we wish to include in the system is not as easy as it may first seem, and has a lot to do with the desired granularity. This word will come back time and time again, and represents the perceived level of detail that a representation manages.

For example, we might wish to define data structures representing chairs in our system. If all the chairs are the same, then we need only represent the standard generic chair, possibly allowing an attribute to identify the color of the chair, but otherwise ignoring the possibility that other chairs exist.

However, the moment that we need to start customizing the chair further—allowing a swivel base on wheels rather than just four legs—we need to introduce another layer of abstraction; we have increased the granularity of the chair definition by allowing the possibility of altering the appearance of the generic chair.

Selecting the right level of detail for the abstraction of the system that is to be modeled is one of the aspects of data-driven designs that is shared with all other paradigms for representing systems. It will often prove to be one of the most difficult steps in the design process.

Process-Oriented Design

The companion to the data structure design paradigm, and often considered a logical consequence of it is the process-oriented paradigm in which it is considered that control is passed from one subsystem to another, often coupled with the exchange of data. This last, however, is not compulsory, since it is quite possible that designs exist in which each subsystem processes information in isolation, and does not directly pass it to other subsystems, despite the fact that each subsystem may share the information that is contained within the system.

For the purpose of this discussion, we shall assume, however, that the flow of control passes from one process to another via the transfer of information. This does mean, however, that each process needs to be in a position to operate on a piece of data, which may be an artificial constraint for designing systems that do not actually revolve around information. These are rare enough as to be ignored.

Objects and Communication

Between the two extremes of data-oriented and process-oriented design lies a paradigm that combines both philosophies, treating the system as a collection of communicating objects, data, and process combined. The result is a system that is broken down in more or less the same way as the real world—with each object having both attributes and methods designed to manipulate those attributes.

The principle behind OO design is that some of these methods are exposed, and others are not. Exposing a method indicates that the other objects can communicate with the object exposing the method by invoking it to render a service as required.

Some terminology is unique to the world of Object-Oriented Analysis and Design, which we will look at presently, but there are some underlying principles that are best to discuss before we look in detail at the OO design process.

Class: A description of a component of the system.

Instance: An incarnation of an object class.

Properties/Attributes: Specific information defining the appearance of an object.

Methods: Processes defined to perform operations on the data, or access it.

Therefore, each object is self-contained, is responsible for its own data, and exposes specific methods that allow other objects to set, retrieve, and access the data, as well as a set of internal methods required for its own management, and some that require it to perform certain functions. If the reader takes some time to look around, they will see that this is entirely in the nature of the real world.

Objects perform functions that have a real world effect, but it is not necessary to know anything about them, other than how to press the right buttons. Programming with the use of an object repository adheres to much the same principles, which is what makes the Object-Oriented methodology so useful to us in the creation of both small- and large-scale software projects.

OBJECT-ORIENTED DESIGN

Thus, we come to a description of the principles behind the Object-Oriented Design methodology itself.

Encapsulation

This is the underlying principle by which all software created with the OO paradigm needs to adhere. In brief, encapsulation means that each object hides as much from the outside world as possible, principally the data that describes it, and the internal processes that enable it to perform the functions for which it is designed.

The advantage of encapsulation is that programmers wishing to use the object in the future need only know what they need to know in order to use the object, not how it achieves the purpose for which it was created. For example, a file object will probably expose pieces of information via access methods such as its name and size, and data can be passed to it, and retrieved from it using other methods, but a programmer wishing to use the object need not know where or how the data is stored.

Thus, the object can be used for storing data on any media, without changing the primary access methods, and without any impact on the rest of the system, simply by changing the underlying implementation. Not only does this mean that the system can potentially support different kinds of media, but that, by adding to the object, it can be reused in other systems such as embedded MP3 player software that needs to access other types of media, such as memory sticks.

It is important, when designing the objects that are required in the system to look at those pieces of information that need to be hidden, and those that are required to be exposed, either directly or via a data access method. There are advantages and disadvantages associated with each technique; usually it is best to follow the principle that all data should be accessed by reference, and not directly. We shall see why when we come to discuss Object-Oriented programming.

Defining the Objects

When preparing a system design using the OO paradigm, the first task that needs to be done is breaking the entire system down into manageable chunks, or objects, of which each fulfills a specific part of the functionality that the system is supposed to offer.

Some of these objects will be parallel implementations of real-world objects, such as objects to control external peripherals such as printers, or virtual objects, such as files, which are abstractions of real-world objects—such as paper documents and filing cabinets.

The principle of abstraction is as much a general software design principle as it is an underlying cornerstone of OO design and programming. It can be an advantage to take this to its logical conclusion for certain systems, and try to create object definitions that mimic the various interactions that would be required in the real world to achieve the same effect manually.

Broadly speaking, three types of systems will need to be defined:

- Those that replace existing manual procedures
- Those that address entirely new problems
- Those that are a consequence of advances in work practices

To design the first type of system, we need only to create a design that is an abstraction of the existing manual system, with objects that more or less correspond to those that already exist. The second kind of system is a consequence of the existence of computers, and includes entertainment software, and applications that would not exist were it not for the creation of computers and other machinery in the first place—such as fractals.

The final broad category includes systems that are alterations of existing ones that result from changes in how tasks are performed in the workplace, and may or may not have a real-world parallel, and, as in the previous category description, it might prove useful to try to imagine how the real-world solution might be arranged in order to create a workable abstract design.

This might prove to be a good approach because of the fact that the OO paradigm attempts to model the real world, and it is sometimes easier to try to think how the same effect might be achieved in the real world and then model the result.

Granularity

So far, we have spoken about objects in a very vague way, without actually going into any detail as to exactly what constitutes an object. The reason is that this will depend on the chosen granularity of each object that is required in the system.

In the previous section, we looked at modeling peripherals such as printers. We might decide that our system can simply contain a printer object, which we can hand virtual documents in the hope that real documents will come out of a printer connected to the system in some way.

Of course, if we have asked a team to create a printer object, or have found one in the object repository that fits the requirements that we have in our design, then the original developer will have looked at the printer object as a system all on its own, potentially consisting of many different objects, either as abstractions (print device, paper feeder, etc.) or as models of the entire printer.

The developers will have chosen an approach that fits their own view of the system—perhaps it is required as a tool to monitor a printer, and as such will need to have objects such as print heads, paper rollers, and so forth. However, they might simply have wanted a generic abstraction of the printer, in which it only needs to be able to inform the system whether a job has printed, whether there is any ink left, and the status of the paper, in which case the object will consist of entirely abstract subobjects.

All of these discussions relate to the granularity of the system. There may even be several designs, each breaking the system down into smaller and smaller chunks, as the granularity increases. Choosing the right point to stop breaking down the objects into smaller and smaller systems is an important facet to bear in mind, because if it is forgotten, the resulting system will be overcomplicated and impossible to realize, or not provide enough control.

Aggregation

This concept of breaking down the system in various different ways depending on the desired level of detail, or granularity, is coupled with one of two profound techniques that we associated with OO design—aggregation (the other being inheritance, which we will come to later). These two concepts are to do with the logical relationship between objects in the system model.

Objects have several different levels of relationship in OO methodologies. The two main views are at the instance level and class level. The instance level relationship between objects is defined by their interaction, while the class level relationship is governed by their design time properties of aggregation or inheritance.

Of course, there is always the third possibility—that two objects have no relationship whatsoever, they merely coexist, offering distinct, unrelated, but important services to the system as a whole.

Traditional aggregation represents a sibling relationship in which the objects share some aspect of behavior or attributes. In this context, we will also add the relationship of sameness with respect to the hierarchical level at which they exist, and we will call this new relationship *contextual aggregation*, since it relates to the relative context of the objects with respect to either their parent or the system as a whole.

Two objects may be considered to be in contextual aggregation if they provide services to a greater system at the same level, without a direct parent that uses services from each. If this is the case, then the objects can be said to be in *direct aggregation*, although they may not, at this point, actually share information or behavior. This last form of aggregation we will call, simply, *aggregation*, which is the traditional definition.

Inheritance

The twin to aggregation is inheritance, and this applies exclusively to the design time view of the system, since in the actual instance view the relationship information encoded within the inheritance tree becomes redundant as actual instances of the inherited objects are instantiated. The reasoning behind this will become clearer as we progress.

The idea behind inheritance is that it creates a network of super- and subclasses that have the capability to share both properties and behavior, overriding them as necessary to make classes that differ in detail, but are largely the same in substance.

The typical textbook attempt to define inheritance invites the reader to imagine different species of animals, and the ways in which they are subdivided and how similar or dissimilar members of the same branch can be.

For example, suppose that we create a class Quadruped, which we intend to be the starting point for an entire family of beasts, ranging from horses to elephants, cats, dogs, and mice. The Quadruped class, then, might specify that members will have four legs and a head that should be on the end of a neck of varying length, plus a variable-length tail.

Then, we might create a subclass, Elephant, which inherits from the Quadruped class, but adds a trunk (possibly as a directly aggregated class Nose as part of Head), specifies a neck of short length, a rudimentary tail, and gives further elaboration on the style of leg.

We can also create a contextually aggregated class, with relation to Elephant, Horse, which also inherits from Quadruped, and modifies it according to examples of the mammal horse. This can go on for each possible variation of Quadruped, almost ad infinitum.

The point is that inheritance allows us to create object classes of similar appearance, all stemming from one parent that groups together their commonalities, which saves a lot of work. There are two caveats, however.

The first is knowing when a child has become so different from the parent that it merits its own class entirely, and the second is being able to group together enough common points between objects that are required in the system to make the creation of a class worthwhile. However, bear both in mind, and the OO design paradigm becomes a useful approach to system design.

Diagramming and Notation

All of the preceding is useless without a good way to get it down on paper, largely because even the simplest system is going to quickly become too complex to hold in one's mind. Therefore, we need some form of notation with which we can create diagrams to show the objects in the system, their relationship, and where they have enough common points to be able to create a network of classes that can be used to define them.

There have been many, many attempts to create a single notation for the description of OO designs, with key work being done by Booch, Rumbaugh, and Jackson in working out methodologies and diagrammatic representations of them. There is even a Unified Modeling Language (UML) that tries to capture the essence of all available paradigms in one graphically rich system.

While this is admirable, and very useful in the design of large, complex, scientific systems, it does not scale very well, and so we are going to keep matters very simple. Our proposed diagrammatic form will consist of:

- Class description diagrams
- Class networks
- Class interaction diagrams
- Object interaction diagrams
- Method definitions

We also need to bear in mind that, since these diagrams are annotated with text, there will be times where a specific property needs to be looked up in the Data Dictionary that forms part of the Functional Specification. We make no provision here for the representation of these properties, beyond that given previously in the *Data Structure Diagrams* section of this chapter.

Figure 9.3 shows a basic class description diagram, and an example of its use.

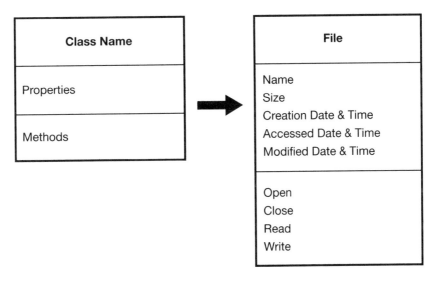

FIGURE 9.3 Class description diagram.

The emphasis, as with any system for diagrammatic representation, is that it needs to be simple enough to create easily with standard office tools, since we want it to be able to scale down in keeping with the aims of this book to serve both commercial software houses and those who find themselves doing development as part of their everyday job. Hence, the simplistic nature of Figure 9.3, created with Microsoft Word.

Sometimes referred to as a type hierarchy, a class network shows the relationship between different classes in the system design. In fact, a class network is more than a type hierarchy, since it shows the aggregations and direct inheritance paths between objects in the system. We are grouping this information in a single diagram, again to respect the aims of this book. Hence, we arrive at a notation such as that in Figure 9.4, again created with standard office tools.

While class network diagrams show how classes are related in terms of their inheritance and aggregation paths, we also need some way of showing the other relationships that are possible, such as the *has* a relationship—as opposed to the *is* a relationship—supported by the class network diagram.

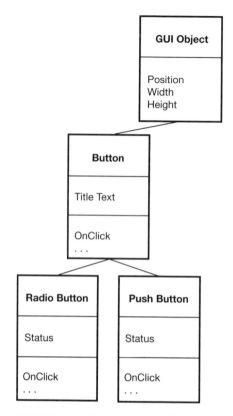

FIGURE 9.4 Class network diagram.

We have two ways of doing this. We can group object names together and enclose them in a bow that denotes the grouping into a superclass or self-contained system, as in Figure 9.5, or we can create a tree that shows how the classes are related as in Figure 9.6.

Figure 9.5 has the disadvantage that it requires many different levels of diagram, resulting in many cross-referenced diagrams that may span many pages. Figure 9.6 dispenses with this by creating a single tree for each level of granularity that the system design requires.

The reader should note that not every possible object or design level interaction is shown in Figures 9.5 and 9.6, just enough to give an idea of how the two different diagramming styles work in practice. In addition, the cardinality of relationships has not been shown in Figure 9.6, where we could usefully implement a system similar to that discussed earlier in the chapter to show that, for example, an Inventory System contains many delivery records, but that each delivery may only be destined for a single address.

FIGURE 9.5 Class grouping.

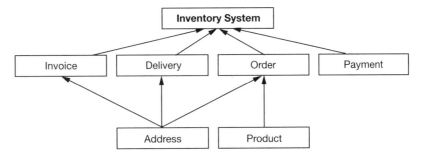

FIGURE 9.6 Class tree.

There is another kind of class interaction that we have mentioned before—the object-level interaction—the various things that can happen when the system and its objects are instantiated for the first time. Figure 9.7 shows an object interaction diagram.

The various interactions depicted in Figure 9.7 are labeled with method names that need to be taken from the final kind of diagrams: Method Definitions. These are merely pieces of documentation that show the same level of definition as in the Functional Specification, and indicate input, output data, and interaction with private methods encapsulated by the class. A Method Definition need be no more complex than:

Class Name.Method

    Uses : <properties from class definition>

    Returns : <if anything>

    Input : <external data>

    Description : <functionality offered>

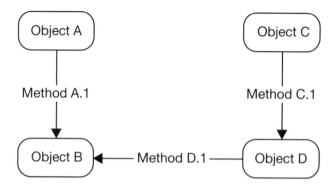

FIGURE 9.7   Object interaction diagram.

Therefore, a basic Web page counter might have a method increment:

WebPageCounter.increment
  Uses: counter_value
  Returns: new_counter_value
  Input: nothing
  Description: increments counter_value, stores result in
  new_counter_value

The basic format can, of course, be refined by using a pseudo-programming language in the Description part of the definition.

OBJECT-ORIENTED PROGRAMMING

One key advantage of using the OO paradigm is that the result of the design process is conceptually closer to the actual programming than with other paradigms. The move from design to implementation requires less of a change in mindset, thus mistakes are less likely to be made in the programming phase of the project.

LANGUAGES

A variety of different languages can be used to implement an object-oriented design, some that were designed from the ground up to be OO in nature, and others that are the result of trying to shoehorn OO style into a procedural language.

C++ is the most popular of the OO languages, although others may not be far behind by the time you read this. Part of the reason why is that C++ has been around a long time, evolving as it did out of the C programming language during the years from 1979 to 1983. At that time, it was called C with Classes, which should give some clues as to what the basic difference between C and C++ is; there is a full class interface, including inheritance, aggregation, and encapsulation.

```
class cWebCounter
{
  private:

    int nCurrentValue;

  public:
    cWebCounter() { this->nCurrentValue = 0; }

    int GetCurrentValue()
      { return this->nCurrentValue; }

    void SetValue(int nValue)
      { this->nCurrentValue = nValue; }

    int Increment() { return this->nCurrentValue++; }
};
```

Anything inside the private section of the class is hidden from view, and the only methods that can be invoked are those in the public section. The cWebCounter method is special in that it is invoked when the object is instantiated. A sister function, not shown here, which would be called ~cWebCounter, handles the cleanup of the object when it is destroyed.

Java was created by engineers at the hardware and systems manufacturer, Sun, to be a simple version of C++, but with full object orientation, and the benefits of being as architecture neutral as possible while still offering advanced features such as multithreading.

Whereas C++ is effectively C with the *class* keyword added (in 1983, that is; since then many more features have been added), Java was created as an object-oriented language from the ground up. Having said that, it does share many key-words and principles with C++, making the transition from one to the other fairly painless for most C++ programmers.

Here is our cWebCounter class rewritten in Java:

```
public class jWebCounter
{
  int nCurrentValue;

  public jWebCounter() { this.nCurrentValue = 0; }

  public int GetCurrentValue()
    { return nCurrentValue; }

  public SetValue(int nValue)
    { nCurrentValue = nValue; }

  public int Increment() { return nCurrentValue++; }
};
```

The reader should agree that this is very C++ in its approach. Where Java differs is that everything is an object, and it is impossible to write procedural code. If one wants to write a program in the traditional sense, that starts at line 1 and runs until line 100, it is necessary to bundle up lines 1 to 100 and put them in a public class constructor.

Objects vs. Modules

Any reader who has had the pleasure of studying different types of programming languages will be aware that there is a branch of the family tree that contains languages such as Object Pascal™, the language of choice for the Borland Delphi development environment, and its relations Modula-2 and Modula-3, the brain children to some extent of Niklaus Wirth.

Such languages are, like C++ and Java, procedural in nature, but with extensions that allow a certain level of packaging beyond the familiar, rudimentary, facilities offered by other procedural languages such as BASIC. Let us start at the beginning, and look at how a language such as BASIC separates code segments:

PROCEDURE: A named piece of code that can be called, with execution restarting on the line following the procedure call.

GOSUB: Causes execution to continue at a named line number, returns once the code block has been executed.

GOTO: Causes execution to jump to a named line number, with no return.

These keywords can be seen as a first attempt to create software that is slightly easier to manage than having one big block of procedural code that must execute from start to end to achieve the correct result. However, we can see the intrinsic problems. The GOTO statement, for example, sends execution off to a specific line number, thereby giving no clue in the code as to what the destination block of code actually does.

This is solved to a certain extent by use of the PROCEDURE keyword, which is essentially a named GOSUB. Even there, however, there are problems in that it merely packages locally; it is no use for creating software that follows object-oriented principles since all the code will be self-referential, or as one of my University lecturers put it—spaghetti code, jumping all over the place.

One side effect of spaghetti code is that it restricts the size and complexity of software, since it is not possible to talk in terms of specific objects having responsibilities without knowing how they work inside. By a similar token, it was not a problem at the time that languages such as BASIC existed, since there were not enough programmers that large teams would be used to work on a product, and the target systems were not powerful enough to do anything much more complex than could be imagined by a single programmer.

Since then, the rules have changed. Hardware allows for larger, more complex software to be created, which also means that larger and larger teams are needed to actually create it. By default, this means that different teams will have different responsibilities, and it is a good idea if they can all work on their own pieces of code (their objects) without disturbing the rest of the application.

Modula-2 and Pascal dealt with this by allowing multiple files to contribute to the final program, and containing pieces of code in MODULEs. Without going into too much detail, a MODULE is like a class in that it is a self-contained piece of code that can be developed in isolation, but which other pieces of code can know how to interface with without knowing how it provides the services it is defined to.

A MODULE, however, is not a class. There is little idea of encapsulation, it is more or less transparent, and instances of modules cannot be created, in the same way that several instances of a given C++ or Java class can exist by virtue of the fact that a class is just a description, and not an actual implementation of a living object.

Modula-3 retains the procedural and textual niceties of Modula-2, and adds a layer of functionality called the GENERIC MODULE that is a nod toward a more object-oriented programming language. It will, however, probably never compete with C++ or Java, anywhere other than the classroom, where it remains an excellent teaching language.

OBJECT TESTING

One of the key advantages of using the OO paradigm is that it makes testing a much easier proposition. Since each object is self-contained, and links itself into the system by exposing various interfaces, once it is tested, we can be sure that it will behave in a predictable and acceptable manner whatever the situation.

This being the case, we may have much more confidence in a system that comprises a number of completely tested, interacting objects. However, we must also be sure that we have tested the objects in a way that ensures that they are completely exercised, and react correctly in all possible situations.

Interfaces

Whichever language or implementation is used, all objects will have a common set of exposed methods, which can be grouped under several different headings:

- Construction and Destruction
- Input and Output
- Action/Reaction

These exposed methods are easily tested, and we will refer to them here as interface methods, or interfaces for short. The first set deals with creating an instance of the object and setting up some basic features, such as any internal variables that can be accessed using the second set of interfaces, which deals with setting and obtaining values for internal (private) class members.

The final set of interfaces provides the basis for communication between objects, and causes the object to perform a specific set of actions, either intrinsically (without stimuli beyond the invocation of the method itself) or reactively, using additional data passed in through parameters.

Such interfaces can be tested by passing a variety of information to them in a way that is designed to exercise the storage and retrieval functionality of the object. This is, of course, particularly of interest to those classes that represent some kind of data storage, such as a string. We could define a string class that looked something like the following:

```
class CMyString
{
  private : // Internal data representation
  long    lLength;
  char * szData;
```

```
public: // Interface code
  CMyString();
    // Constructor #1—make an empty string

  CMyString(long length);
    // Constructor #2—make a string of length

  CMyString(char * data);
    // Constructor #3—build from a constant

 ~CMyString();
    // Destructor—called when instance is deleted

  char * GetString();
    // Should return ->szData or NULL

  long   GetStringLength();
    // Should return ->lLength, 0 or -1, depending

  long   SetString(char * data);
    // Put new data in

  long   ResizeString(long new_length);
    // Extend/reduce to new_length
};
```

The interface members in the public section of the class definition make it reasonably obvious how we could test the implementation. Since we can know the values that we put into the system, we can verify them against the values that we retrieve, using the relevant interface member function. Should an error occur, or the data that is retrieved differ from that which we supplied, then we know there is a problem with the implementation.

We can test on each of the three levels that we looked at previously—at the construction level, at the "get and set" level, and finally, at the interaction level. Each level has a slightly different kind of effect on the internal data structures, and moving from one phase to another ensures that, once the final area of functionality has been verified, we can say that the object has been fully tested.

Therefore, the first level of testing will involve creating and destroying an instance of the object, in as many ways as possible. For a class that handles memory blocks in some way, as the CMyString class in the previous code sample does, it is important to test all the core values to verify that the object can correctly handle large and small strings.

We then need to verify that the data access interface members work correctly. We do this by initializing the object with a given set of values, then retrieving them to verify that the correct values have been set, before altering them in some way, and testing for effect.

Finally, we have to test any functions that have an indirect effect on the data stored inside the object, and then check that the correct behavior is exhibited by retrieving the data. The first two areas of testing are reasonably easy to test, since we can do a direct comparison on the data that has been fed into the object, as well as that which we have retrieved. The following code sample shows how the first two areas can be tested, by using a simple test case of a standard string for the CMyString class.

```
#define STRING_30 "a test string of 30 characters"
#define STRING_30_LENGTH 30

// Test string initialized with no data

    CMyString * oString = new CMyString();

    oString->SetString(STRING_30_LENGTH);

    if (oString->GetStringLength() != STRING_30_LENGTH)
      // Report error

    oString->SetString(STRING_30);

    if (strcmp(oString->GetString(), STRING_30) != 0)
      // Report error

    delete oString;

    // Test string initialized with a length

    CMyString * oString =
      new CMyString(STRING_30_LENGTH);

    oString->SetString(STRING_30);

    if (oString->GetStringLength() != STRING_30_LENGTH)
      // Report error
```

```
if (strcmp(oString->GetString(), STRING_30) != 0)
  // Report error

delete oString;

// This is NOT AN EXHAUSTIVE LIST OF TEST CASES
```

As you can see, although the idea of testing objects using their interfaces is not a difficult concept, it can become quite involved. The third area of testing is slightly different, as it requires the developer to think up test cases that adequately manifest themselves through changes to the underlying data such that it can be retrieved in order to verify that the correct change has indeed taken place.

We can consider object testing as being on the same level as unit testing, since once the object is fully tested, it can be handed over for integration, either to a module or to the creation of the final system, depending on the facilities that it offers.

SUMMARY

It should be reasonably obvious that this chapter has been more about showing how the OO paradigm stands far above the others in terms of fitness for use as a means to develop complex software systems. It is also scalable such that even a reasonably small system can benefit from OO principles to yield a better quality product.

We have also shown how the points that make the OO paradigm so desirable can be best exploited to improve the quality of the end system. Essentially, if each system is made up of objects, we can take advantage of:

- Separate development teams
- Reusable source code
- Ease of testing
- Conceptually closer to real-world thought process

These four advantages alone make enough of a case for using the OO paradigm wherever possible.

10 Reusable Code Guidelines

INTRODUCTION

We have been using the Object-Oriented (OO) paradigm as a basis for establishing good analysis, design, and development practices, but the true benefits will only be realized if guidelines are set out for code reuse. OO design is built upon principles that facilitate code reuse, but it is not automatic; simply applying the principles will not lead to code that is reusable by default.

The benefits of reusable code are clear—objects that are created to solve a particular problem can be used again whenever the same problem needs to be solved in future projects. The chances of this happening are quite high, since development tends to happen within specific niches of the market as the company builds up expertise in a given market area.

In this chapter, we will examine the problem from the ground up, using the established organizational model to facilitate good object-oriented design leading to high-quality reusable code. These objects and components that result from code reuse can also be sold as products in their own right; and of course the reverse is also true—good quality components and objects can be brought in to facilitate development of a particular product.

In fact, it is quite feasible to create a product that is entirely a result of gluing high-quality reused components together, which is another way in which software development is different from any other industry. Code components do not wear out, are cheap (free) to duplicate, and can be combined almost effortlessly together—as long as the initial design is correct, of course.

REUSE AS A POLICY

To make use of the advantages of object reuse, it is necessary to identify the process as part of official company policy. That is to say, the system architects, designers, and programmers all need to approach their tasks with the view that the code they develop may be reused at some point in the future, and should therefore adhere to specific principles. Anything that the company produces in the process of creating software can be reused; from design and documentation to the actual code itself.

The reuse process itself should always be based on the decision process to create new code when nothing else exists that can be reused. This means that writing new code becomes a last resort, having already followed through the process of:

- Finding existing code
- Adapting code that represents a close match to the requirements
- Establishing no prior technology in existence

Following these priorities will have two effects. The first is that less code will be written, less mistakes made, and the overall quality of code in the code base will improve. The second is that time to develop will be reduced, manifesting itself in a direct increase in efficiency.

Documentation

The key to successful object reuse is having a supporting framework that standardizes the identification and description of objects. This serves two goals:

- Ease of object creation and maintenance
- Ease of object location and use

Essentially, by accurately defining the information that must be present to lodge an object in the library, it will be easier to find and reuse, and will follow strict guidelines for design and development as laid out in the supporting documentation.

Of course, the source code itself is also regarded as documentation—being plain text it is also possible to index the source code in such a way as to provide a clue as to what functionality the source provides. The clues will be given by function and variable names, as well as the (hopefully standardized) comments that are included in the code that detail what each function is supposed to do.

Object Description Document

Each object needs to be accompanied by some form of documentation that includes information about what the object is supposed to offer as features, or services. In this case, when we refer to "object" we are using it as a basket term to cover:

- Internally created source code objects
- Open Source acquired source code
- Third-party maintained source code

We are not using it to cover those pieces of code that are closed source, binaries that have been acquired as part of the Component Gallery. These will follow a slightly different pattern of reuse, as we cover later in this chapter.

The Object Description Document is used as a way to identify the key services that the object offers, in terms of mapping inputs to outputs (data or actions, in both cases), and identifies it in terms of source code language, operating system, and any additional constraints that might be relevant.

It can be partially derived from the design or specification documents that led to its creation, and as such, it is possible that the Librarian or a staff technical writer will be able to produce these documents. It is important that they all follow the same kind of layout, and that this layout is well published throughout the organization.

Of course the Librarian function may not be an existing staff position, but, as we discussed in Chapter 1, "The Liaison Center," an amalgamation of a general agreement to manage code in a certain way, backed up by sufficient tools to make the process workable and efficient will naturally result in a set of documents that are proof of the procedure.

Furthermore, the documents need to be searchable, and refer to the real location of the source code, since they will probably not be stored in the same place. There will probably be a need to add some kind of access point that extends beyond the simple operating system. One of the most effective ways of doing this might be to convert all the documents to a read-only format (such as Adobe PDF), and create a mini Web site that provides a search function and a modicum of organizational logic behind the storage of the documentation.

Using Web technologies will mean that the front end (user interface) can be easily created and maintained, and that users do not have to wade through directories on unrelated files just to find the folder that contains the documents they were looking for.

Extending this to store everything in a backend database is probably only relevant if the object collection is destined to grow to a substantial size. At this point, the database should be created with a form that follows the document headings and information. It will lead to a faster search of object information, and ultimately a much easier Web programming task, at the expense of introducing another piece of third-party software into the chain.

This third-party software will need maintaining, monitoring, and possibly training before it can be used, but the end result might just be worth the additional effort, depending on the goals of the organization in question.

The Object Description Document (or Table) should be split into three sections. The first should cover general identification, such as project type (Graphics, System, Database, World Wide Web, etc.), language, and operating system.

The second section defines the input to output mapping, either data or actions, along with a description of purpose. It is essentially comprised of free text areas that allow the technical writer to reduce the specification or design of the object to a description that will allow a third party to decide whether the object is going to satisfy their requirements.

The final section comprises keywords that underline the technologies used and the key areas of functionality. For example, we might have a sorting object, capable of sorting integer values. The general identification section might look like:

> Project Type: System
> Language: C++
> Operating System: Independent
> Object Type: Library

The descriptive text might then read:

> Input: Array of integers, with size information
> Output: Sorted array, replacing original
> Description: Sorts the array using one of a selection of algorithms

The keywords for this object might be:

Keywords: integer sort, bubble sort, quick sort, divide and conquer

Implemented as a Web solution, the users might be confronted with a Web page that allowed them to select the operating system and language, for example, and some keywords to define what it was they were looking for. Matches should then be returned from the system to allow the users to select the code that most suits their needs. A good example of this can be seen on the Source Forge Web site (*www.sf.net*).

While this might force programmers to properly think through their solution prior to its creation, and as part of the coding process, most programmers tend to think in terms of writing, testing, and rewriting their code to match the functionality required.

In such cases, it is perfectly possible that the documentation referred to here can be extracted from the comments that are present in the code, which removes the manual procedure to create a document that covers the salient points, and replaces it with an automatic function based on the extraction of commented information.

For this to be a workable solution, however, it is necessary to entirely standardize the commenting process such that they become a partial replacement for the documentation. However, it should be pointed out that the emphasis on producing the documents does help to ensure that the process is being followed correctly, which will usually lead to an increase in quality.

Source Code Documentation

We mentioned that the source code is in itself documentation. A simple text search of all files with a given extension denoting their programming language would probably be enough to give someone with a considerable amount of time to find the exact routine that suits their needs. It is, however, not very efficient.

However, we can make source code much easier to search by introducing very simple concepts, such as tags. For example, the keywords section from the previous object documentation example:

Keywords: integer sort, bubble sort, quick sort, divide and conquer

This gives us some clues as to what the functions inside the object might offer. Suppose that we want to create a string storing library, and, knowing that one does not exist, but not wishing to reinvent the wheel, we decide to look for a sorting library, and reuse some of its code.

Now, this means that we need to access the object at a different level, because we want to look at individual functions. The code may not be part of an actual sorting library; it may be an offshoot of another library offering a different service, coupled with which, we might also like some additional string handling functions.

When the author of the integer-sorting library created the source code, he might have used simple tagging, such as:

// Input: array of integers

// Purpose: sort

When we want to find a function that handles arrays of integers, we can then look for a text string "// Input: array of integers". The "//" is the C language way of specifying a comment line, one that will be ignored by the compiler. We are using it in a very specific way, to denote the start of a tag line, and we have established a standard for specifying a tag:

// <tag>: value

It is important to note that there is a single space between the // and the tag, and that there is no space between the tag and the :, and one space between the : and the value. These are important because they allow a simple text search to pick up the unique tags without requiring us to implement specific logic that can read (parse) the C code file.

The set of tags that is permitted needs to be specified, as does the range of possible values for each tag. These need to be controlled such that, if a programmer introduces a new value for a tag, or wants to introduce a new tag, the new properties are added to the document that covers such matters, and the change is passed on to the users of the system.

In this way, we can be sure that the set of properties remains as rich as possible, and that everyone knows what it is. This is necessary so that they can use the most appropriate information for their source code, when creating it and when they want to search for it.

Process Documentation

There also needs to be a policy document that describes how the objects can be located, how they are to be obtained and used, and what the feedback procedure is for returning errors to the original authors, or fixed errors back to the code base to enrich the object repository.

We will cover exactly what these two entail in the next two sections, but it is important to note that part of the overall documentation has to cover how the entire system is to be used. It avoids staff members accessing the object repository in an inappropriate fashion—either intentionally or through laziness or negligence.

There should be a channel through which all communication passes, to the effect that, whatever change management system is being used, the process by which the code is obtained, once the relevant object has been identified, does not conflict with it. After all, the object repository is a service that should be easy to use, and yet be subject to guidelines that will ensure that its integrity remains intact.

Searching and Using

Once the source code has been located, it can be used. There are a variety of ways in which a specific object might be used, depending on the kind of object that has been located. For example, if it is just source code, designed to be integrated with another project, then it must be extracted as source, melded with the existing code, and recompiled.

However, if it is a library, then there is every possibility that it can just be lifted and used—especially if it is a run-time library (such as Windows DLLs or Linux shared object libraries)—in which case, there is very little intervention required.

Source Code

Requests to reuse source code need to result in a set of actions that ensure that its integrity is retained, on the one hand, and that any changes that are introduced from other sources, as a result of feedback, are offered to the users of that source code.

This is a nontrivial exercise, and will require source code control tools and techniques that are beyond the scope of this discussion. We do touch upon it briefly in other chapters, but you should note that expert consultation will be required to ensure that the correct solution is used.

The theory is that the source code needs first to be lodged in the repository, and it then becomes stable, as a release. If the code is not fully tested, contains errors, or does not fulfill criteria laid down by the policies of the object repository, it should not be included, or offered for use.

Upon use, the source code takes on a new lease on life within someone else's project, but the original code remains in the repository. This is effectively a branch of the code base, which is dealt with by most source code control systems in a manner that is efficient and traceable, which is important.

Remembering that the source code is either a stand-alone object, which is designed to be reused in its entirety, or a piece of fully tested source code that is not an object but a collection of functions or routines designed to perform specific tasks, then there may come a time when the code has been extended beyond its original scope. The other alternative is that the code has been fixed as a result of locating an error that was not previously found.

In either case, the source code must be combined with the old copy, in order that the changes become part of the object repository. Again, this should be handled transparently by the source code control system. What is important is that every user of the code, for projects that are in development, or maintenance, is informed that the object has been changed, since they may want to incorporate those changes in their own project.

It is up to the organization whether they adopt a strict single copy policy for reused source code. If they do, then they must be aware that there might be repercussions and conflicts when two projects need the same source code at the same time.

Libraries

The alternative to source code is to insist that all objects are made available as libraries, be they statically linked (at compile time) or dynamically (at run time). This removes any complex source code management problems of having multiple copies of the code floating around, since only the object or executable code is made available to the project teams who want to use it.

Subsequently, the source code control task, which is the responsibility of the Librarian, is much simplified, but at the expense of slightly reduced reuse flexibility. The advantages are numerous, and using pure libraries may even contribute to higher quality software, since the feedback, reporting, and adapting procedure will follow formal change control lines.

Otherwise, the process is essentially the same as for source code. The users search through the repository, and, finding a suitable candidate, locate the library. They should register the fact that the library is being reused with the Librarian so that they can be kept appraised of any changes to the code that might be to their advantage.

The big difference is that this is where the user's obligations end—the library is out in the wild, but since they cannot change it, it will never return to the repository. There is no need for it to. If the users find a problem or want to change the object, then they need to go through a separate procedure, which we discuss in the *Feedback* section later in the chapter.

There is one further caveat—platform. Unless the developer of the object goes to lengths to make sure that it is available in a number of different platform flavors, then only the platform for which it was compiled will be able to support it. Since we are not allowing the user access to the source code, except under certain circumstances, there is nothing that can be done about this, except request a version that is targeted for a different operating system.

There is a line of thought that states, in no uncertain terms, that nonportable, or platform-dependent, code is not good quality code. However, in many cases, there will be code, particularly that which deals with the operating system, GUI, or devices that will be inherently nonportable.

This problem only evaporates if the target platform is always one for which a virtual machine is always available, such as is the case with Java applications. If this cannot be relied on, a certain amount of code will be produced that is platform dependent. The effect of this can be mitigated by attempting to keep it as separate as possible from the main code base that performs the core functionality of the product.

Tools

Finally, the repository may contain a number of specific tools and utilities that are designed for reuse. These will follow much the same pattern of reuse as libraries, in that they cannot be changed, except under specific circumstances, detailed later in this chapter, but differ in that they do not become part of the new project, but are additions to it.

Feedback

Part of the point of actually having an object repository is being able to feed changes back into it that will enrich the collection, and to ensure that there are always objects available that can be used in a variety of different situations.

On the one hand, there is the addition of new code, be it new objects, or changes to existing ones, and on the other, corrections to improve the quality of the objects offered to the project teams.

To achieve this Utopian view of the object repository, it is necessary to formalize the feedback procedures so that the Librarian (be it automated or manual) has an easier task in managing updates, removals, and the communication of feedback issues to all parties concerned.

The way in which this is done will change depending on the type of object that is being reused, and the Librarian philosophy. A purely automated system in which the source code management tool is fully integrated with the software development infrastructure could feasibly enforce commenting upon updates, and circulate the information accordingly.

However, if we discuss it as a manual process within the course of this book, it becomes clearer as to what function is being fulfilled, as well as offering a manual solution for those organizations where an investment in tools and training represents more of an outlay than designating responsibility to an individual or collectively.

Source Code

We mentioned that one of the problems with allowing direct object reuse at the source code level was that it became more difficult to deal with issues surrounding source code control. The problem is that if we release the source code, the user may change it, and want to return that change back to the repository, leaving the Librarian with the unenviable task of trying to establish what has changed, and how to write it up so that everyone can take advantage of the new features.

This can be automated to a certain extent by the source code control system; however, it will not be intelligent enough to identify the nature of the change, only that a change has taken place. Commenting in the code will help to establish exactly what the purpose behind the changes was, but this may not be adequate.

Consequently, any source code reuse scheme needs to have a feedback procedure that clearly identifies the changes that have been made, and the reason behind them, as well as information regarding testing and verification procedures that have been followed. Without these, it will quickly become very difficult to manage the change process, and the entire repository might be at risk.

It might be a more sensible approach to disallow source code changes unless they have first been agreed to by the Librarian. Essentially, this will mean that only source code that has had its design accepted will be allowed into the repository. If a change needs to be made, whether it is to correct an error or as part of the evolution of the object, then that must first be agreed to before the new code is admitted.

In an automated environment that should be within reach of even small development centers, using a set of tools that can perform the source code management, and feature and error tracking in conjunction with the Liaison Center concept can replace the Librarian and Source Code Manager completely.

This does not change the aim of the preceding discussion—being able to accurately track and manage the source code and quality thereof—but it should make the process more manageable. At the end of the day, however, the quality of the code and its general fitness for use will usually be enhanced by the manual application of standards by a member of staff.

Libraries and Tools

Restricting the object repository to libraries is possibly one compromise that will benefit small- to medium-sized organizations. The model is different in that each object becomes a self-contained library—either of functions or a single, useable, object—with the only possible way to change it being via a formal change request.

This approach benefits in all areas. It helps to promote the idea of a customer-oriented environment, which is key in software quality control, as we will see later, and provides an audit trail that can be followed to keep track of issues surrounding the quality of delivered objects.

It does lack flexibility, and can be less efficient than direct source code sharing, but in cases where a good balance between quality, flexibility, and efficiency is required, it does present something of a balanced approach.

The key is in providing feedback via Change Request or Problem Report mechanisms. On the one hand, requesting that a change be made to extend the functionality offered, and on the other, reporting an error in the hope that it will be repaired in a timely manner.

The latter will need to have severities attached to it such that the more severe errors that are preventing progress in a given project are corrected as fast as possible. There is also the possibility that if the original programmer is no longer responsible for that object, that the source code correction can be performed by the requester—however, this will mean that the source code control system has to be aware of what is going on.

In fact, it will be a far simpler approach than direct source code sharing, as only one person will be able to check out the code for changing, and it will usually not branch into another code base. When direct source code sharing is permitted, there is always the danger that the object becomes extended beyond its original scope, and needs to be broken up after the fact.

Any tools that have been created will follow the same pattern of feedback as libraries, in that they become internal products that can be shared and maintained by a team of programmers, who will usually accept change requests and problem reports for as long as they are working on the tools.

At a certain point, however, they will move on to other projects, and the code will become part of the general repository. It then becomes the responsibility of the organization to make the decision as to whether the requester can then change the code, or whether a dedicated team of code maintenance specialists will be required.

This is a specialist position and requires that the staff are able to quickly read other people's code, assimilate the changes needed, and make them accurately. Many programmers will point out that this is usually quite a difficult task.

SYSTEM GRANULARITY

We have spoken a few times in this chapter about policy decisions that the organization needs to make to ensure that the processes that are put in place to promote object reuse work in a manner that is in keeping with both the nature of the projects that the organization carries out and the guiding philosophy of that organization.

One of these policy decisions is linked to the construction of the repository and the kind of artifacts that will be stored there. *System granularity* is a term that we have used a few times and essentially points to the programming style and nature of the projects that the organization carries out.

A fine-grained system will consist of many pieces of borrowed source code, usually individual functions, and possibly objects, either contained within source code blocks or precompiled libraries. A coarse-grained system will generally consist of larger units that are glued together by small chunks of logic. The object repository can also consist of fine-grained system components, coarse-grained ones, or a mixture thereof.

Fine-Grained Repository Artifacts

The finest grain of repository artifact is offered by using code reuse at the source level—we can effectively share code right down to the function level. This implies that we can reuse individual functions that may or may not be part of specific objects.

In theory, we could even reuse smaller units, such as individual code blocks, but this is probably not appropriate within the context of software engineering. The smallest unit that should be reused in most cases, using the Object-Oriented (OO) paradigm for design and implementation, is an object definition: in C++, a class; in Modula-3, a module; a Java class, or an Eiffel package, to give a few specific examples.

In very specific system cases, we might choose to reuse functions, but this will violate some of the principles of OO design and programming, although we can minimize this by using the functions within objects when we come to glue the system together.

The disadvantage of using a fine-grained repository down to the individual function level is also that it becomes much more difficult to extract the functions and feed them back into the repository should they become enhanced or corrected. It also becomes more difficult to know whether it is the functions that are at fault, or the code that is holding them together.

In the end, if fine-grained repository artifacts are to be used, then it is best to do so at a level that provides the eventual user with a set of interfaces and a level of abstraction that encapsulates the actual processing in a reusable package.

If the piece of code that is being reused can be treated as a black box, it becomes much easier to debug, enhance, and feed back into the repository once it has been changed, and all of these are important when trying to practice efficient object reuse.

Coarse-Grained Repository Artifacts

At the other end of the scale, we have the possibility to only reuse objects at the highest level of abstraction. This means directly executable code in the form of statically linked libraries that are built into the executable at compile time, dynamically linked libraries that can be accessed by the software during run time, or executable software applications that are standalone pieces of code in themselves.

This last is usually applicable in an environment where most of the processing is done in an offline, or noninteractive fashion. In this case, we mean noninteractive in the sense that, at the operating system level, there is no direct user interaction between a constantly running application and the operating system.

The most common example that comes to mind is the Web interface offered by most server packages that allows packages to be executed at the request of the individual browsing the site. In such cases, some of the pages will be, at least partially, the result of a piece of software that is executed, and returns HTML or equivalently formatted information that can be displayed by the browser.

The same can also be applied to systems that use back office processes such as large databases, where the database software becomes a very coarse-grained reusable system component that can be interacted with only via its API. The operating system would also fall into this category, although it would overlap with the concept of libraries, since most platforms allow software to interact only via dynamically or statically linked libraries.

Coarse-grained object reuse is important because it offers the most elegant solution to the concept of integrating the OO paradigm with the idea of maintaining a repository of reusable artifacts. It may not always be practical, however, since if the organization is involved in development projects on a variety of platforms, there may be a necessity to support a number of different library formats.

Using the next level down—object code in a standard format—will get around this problem, while keeping the obvious advantages of programming language transparency and total encapsulation of internal data and processing. The objects in a coarse-grained repository can be treated as opaque boxes—we feed them data, and we receive responses, but beyond that, we do not know how they do what they do.

A further advantage is that there is a very much reduced dependency on new code in a project that has been constructed entirely of coarse-grained components. Of course, the components have to exist, and so it is only when the organization becomes more mature that the benefits will truly be reaped.

Object Reuse vs. Component Galleries

The other side of the repository is the Component Gallery. This is a coarse-grained solution, since it will contain only licensed pieces of code such as full libraries, tools, and back office style executables that the organization has the right to redistribute. The components are usually acquired as commercial offerings or through a system of procurement in cases where no obvious solution exists on the market.

They are different from artifacts in the Object Repository in that the organization probably has very little control over their evolution. There may be new features being added by the original authors of the components, and there may be a mechanism in place whereby the organization can report errors and request solutions to those errors, but overall, the process will not be controlled by them.

If the source code to a specific component has been licensed by the organization, it is a wiser move to include the component as part of the Object Repository, and effectively take control of the source for future iterations. However, this may mean that the code base becomes branched, and it might prove difficult to introduce changes into the code base that are recommended by the original authors.

Whether source code becomes part of the Object Repository will probably be dictated by the agreement that the organization has with the authors of the code. If there is no contract that specifies that the code base will be maintained, then there is probably no need to try to ensure that changes can be merged into it.

THE OPEN SOURCE REVOLUTION

Although a good source for reused code is the organization itself, and components can be purchased from commercial sources, there is an alternative: Open Source. Using Open Source code enables a fledgling organization to obtain a large amount of royalty-free, reusable source code without expending large amounts of resources.

The arguments for obtaining code in this way are the same as for reuse in general; there is little point in reinventing the wheel, if someone else has already created the code, and is willing to share it for the good of the community, then it is obvious that a software engineering company using it will have a clear advantage in the marketplace.

However, as we will see, a number of caveats govern the acceptable use of this, otherwise free, code, which relate to a reasonably complex web of licensing agreements, each with diverse consequences for those wanting to make use of code covered by licensing.

Open Source Code

The Open Source movement comprises a collection of talented programmers who release source code into the community in the hope that it will be used, extended, and handed back to that community. The Linux operating system is a classic example of an Open Source project in which the operating system was put together as an alternative to standard Minix and Unix systems and released to the general public.

Not only was the executable software released, but the source code to the entire project was released under a special license that enabled people to copy it, modify it, and use it in any way that they saw fit—except selling it.

The result was that support for Linux grew beyond the initial hardware configuration for which it was created, and continues to be extended as new pieces of hardware appear on the market. If a Linux user has an esoteric piece of hardware that he wants to use with the operating system, then he must either find someone with the same hardware to create a driver for it, or do it himself.

Finding Open Source Code

Open Source code can be found in a number of places. At the time of writing, the key proponents of Open Source are the Source Forge project and the Open Source Development Network. Both of these organizations serve as a distribution point for code, but the Source Forge Web site also provides Open Source projects with Web hosting, change management, and virtual communication tools.

License Types

There are a number of licensing issues related to the use of Open Source code, which stem from the initial intentions of the Free Software Foundation, who are the driving force behind the Open Source movement. In fact, one needs to understand that they are using the term "Free" in a way that is not limited to monetary implications.

Free in this case does not mean that the user need not pay for the use of the source code, but that the distribution model is also free, as in freedom. Part of that freedom is in the protection of the original author of the code, such that subsequent authors can be, by accepting the license agreement and using the code, bound to the same principles under which the code was licensed in the first place.

This is not the place for a detailed discussion of the various license types that are available under the Free Software Foundation Open Source licensing scheme, but it is worth noting that there are some very broad categories into which the most common of these licenses fall.

The three most common licenses that are used in Open Source circles are:

- Public Domain
- GNU GPL
- GNU LGPL

Public Domain licensing is the least restrictive, the GPL is one of the most, with LGPL lying somewhere in between. The licenses were created so that authors of software packages, tools, or source code could take advantage of standard, Open Source documentation that meant that they could be protected while not needing to retain an army of lawyers.

Since the actual text of the licenses is Open Source, people can use it in their agreements of use for their software, in the knowledge that it has been well thought out. Unfortunately, since it is the product of legal professionals, it can also become reasonably difficult to understand, which is why most of the licenses come with a statement of general principles at the start, from which we derive the following discussions.

Public Domain

Anything that is placed in the Public Domain is free for anyone to use, reuse, change, or distribute as they wish, if the original authors are acknowledged. In fact, most authors leave this at the discretion of the subsequent users of their precious code. Effectively, they are sharing the fruits of their labors with the world at large, without hope of remuneration or recognition.

The code can be integrated with other packages, sold, reworked and sold, or simply used as is. The license does not restrict use in any way, and is therefore one of the most open of the Open Source licenses.

GNU General Public License

The GPL is perhaps the most widely used of all the Open Source licenses, and is the license that covers distributions of code such as Linux, and most of the GNU tools, which includes compilers and operating system tools for Linux and most mainstream operating systems.

The GPL also happens to be one of the most restrictive of the licenses, which, if it is applied to the fullest extent effectively prohibits the person using the code that is protected by the GPL from using it directly for commercial gain, without ensuring that the resulting product is bound by the same terms—in other words, if the product contains code covered by the GPL, then it should also be covered by the terms of the GPL, and not a more restrictive license.

If we take the most famous example—the Linux operating system—we can see that this is perhaps, while restrictive, not as restrictive as we might at first think. In fact, a number of companies are currently making money from the distribution of Linux. They are not selling the operating system itself, and the license that accompanies it usually states that the operating system component can be redistributed by the purchasers as they see fit.

What these companies are selling is the repackaging of Linux, including tools for its installation and tools to help the user after installation, and documentation, not the operating system itself. The only reason this is permitted under the terms of the license is that the derived works and original code are freely distributable.

More than this, the source code to the entire project has to be available to the end user, either distributed as part of the package, or upon application or download, in cases where the source code contains some GPL portions. If it does not, and merely sits on top of the GPL code (such as an installation routine created from scratch for Linux), then the source does not need to be released.

What is not acceptable is to sell the package, which includes GPL covered code, and then close the code so that nobody has access to it, as is the case in traditional software distribution models. It is this that makes the GPL a little too restrictive for outright commercial exploitation, under certain circumstances.

The GPL is worded such that it prevents exclusive use and commercial advantage, as it states in the Open Source Initiative Web site FAQ:

> "What you can't do is stop someone else from selling your code as well. That just says that you need to add extra value to your code, by offering service, or printed documentation, or a convenient medium, or a certification mark testifying to its quality."

> —Extract from *www.opensource.org/advocacy/faq.php*

It also means that the software should be passed on to a third party (via sale or gift) under the same terms that the original author extended in the first place. Thus, depending on the exact mix of components, and whether the work was entirely derived from the original, it may mean that all the source code of the new work must be made available to the general public for a period of at least three years.

GNU Lesser (Library) General Public License

Perhaps in recognition of this need to be able to commercialize Open Source code, the LGPL provides a way in which the original author can distribute code freely, while protecting the users from having to disclose their own, proprietary developments. It was specifically designed to allow this by only being applicable to the code that it covers.

Unlike the GPL, which also covers derived code, the LGPL enables developers to use the protected code to their advantage. However, as long as it remains unchanged, and is not used as a base for building proprietary code, the product can be sold, distributed, and fully exploited.

Thus, the code that is protected by the LGPL needs to remain a self-contained opaque library, or at the very least, a compiled-in component that remains unchanged during development. This puts it clearly in the case of being part of the Component Gallery, while the source code might become available.

Depending on the wishes of the author, the exact interpretation of the LGPL that is covering the code, it will probably not be appropriate to try to place that source code in the Object Repository, since it cannot be owned by the organization—only "borrowed."

The Open Source Advantage

The key to understanding the driving force behind Open Source is to treat the arrangement as a way of sharing the fruits of the labor of the entire programming community, for the good of the community. In other words, if you take a product, and make something from it, then perhaps you should extend the possibility for others to do the same with your work.

Using Open Source code may seem slightly cumbersome and more than a little restrictive at times, but it comes with a clear set of advantages that are illustrated by the flagship "product," Linux. In a nutshell, Open Source works because the people who use the code are the authors, or are technically competent to assume responsibility for part of the functionality.

This means that as soon as they isolate a need to extend the operating system, such as the support for a new or exotic piece of hardware, they are usually capable of changing the source code themselves so that this new feature is included. They must then, under the terms of the license, make this customization available to all, which makes sure that the benefits of their work are felt by the entire user community.

In the case of Linux, it is clear that this will lead to an incredibly rich variety of drivers for all manner of esoteric devices and file formats. For example, there are readers and writers for closed binary file formats such as PostScript or Adobe PDF documents, not to mention support for the most cutting-edge graphics and sound cards.

Using the Open Source initiative, the user base for a given piece of software will also grow more than if the software had to be purchased outright as a commercial offering. This leads to wider exposure for the component, and the possibility that the errors will be located and fixed in certain circumstances. This is only made possible by the fact that the source code is made available to the end users, should they wish to read it.

The effect on the industry is that, as long as the licenses are respected, and the Open Source ethics applied, the code sharing that goes on between licensees and the authors of the software can result in a useful dialog that is beneficial to all parties.

In the end, the company may just end up charging the client for the installation of an Open Source solution, rather than having to create the solution from scratch.

Therefore, there is a clear competitive advantage for those who are in a position to level the power of the Open Source model. While they may, on the one hand, try to be incredibly protective about the product or service they are offering, it cannot be denied that a company that has the experience and knowledge to offer customization, installation, and documentation of an existing product can do so at a cost far less than having to build a new product to solve the problem presented to them by the client.

As long as the client has no issue with the use of well-supported and documented Open Source solutions, there is no reason why this should not become part of the organization's approach to software engineering on a corporate scale.

Open Source Ethics

Besides the licenses, the Open Source initiative is also governed by a set of unwritten ethics. They started with the story of stone soup, as told by a company of the same name. The tale is of a man who finds himself in a village, hungry, with nothing but a cooking pot and some water. So, he sets to locating some fairly sizeable stones, and commences to boil them.

A passing villager looks on, and asks the man what he is doing. The man replies that he is making stone soup, and it tastes very good, but would be so much better if it were to have some onions in it. Grudgingly, the villager agrees to provide some onions, in return for some of this amazing stone soup.

Other villagers pass by, each wondering what the man is doing, and each time they get the same answer—the man is making stone soup, and it is the best he has ever tasted, but would be so much better if he could just add some cabbages, tomatoes, and, finally potatoes to the mix.

Once the vegetable (stone) soup is ready, he sits down with the villagers and they eat the wonderful soup together, everyone benefiting from the resources of each other. The message is clear, and it is one of the underlying principles of Open Source.

The licenses were also created with the same philosophy—that, in using them, it is the prerogative to give something back whenever possible. This could be comments for making the code better, features to be added, or corrections to errors.

SUMMARY

This chapter aimed to give the reader an insight into how code can be reused in a manner that is efficient and practical. The level to which it is implemented will, as always, depend on the resources available to the target organization. It is, however, always wise to try to plan larger than is currently needed, in the assumption that the company will grow, rather than standing still.

It also pays to adopt a consistent approach to code sharing, either a fine grained or coarse grained solution, but trying not to mix the two. This may not always be possible, but as a general rule, it will depend on the kind of projects that the company is embarking on.

If they tend to be highly specialized, technical projects, then a finer grained solution may be required, since the proportion of code that can be shared between projects will be lower than if the company is working in a standardized field, such as web development.

Finally, it also pays not to write of Open Source solutions due to the licensing, as so many companies seem to do. It is worth taking the time to examine the marketplace and see what free and open solutions are available, and establish a paradigm for their reuse that benefits the company but does not infringe on either the ethics or the licensing agreement that the company enters into upon their use.

11 The Object and Component Archive

INTRODUCTION

In the previous chapter, we looked at ways in which code reuse can be practiced within the confines of our organizational model laid out in Part I of this book. However, the guidelines on their own offer little practical advice on how to provide for the efficient storage and retrieval of the objects and components that are created as a by-product of performing various software development projects.

This chapter looks at creating an organization-wide Object Repository, and the techniques described could be applied to almost any kind of documentation storage. Indeed, many of the technologies that are currently used in large-scale documentation storage and retrieval projects can be applied to the storage and retrieval of source code. The key is in how the code is described and indexed within the repository.

Again, the material presented here works best with the Object Oriented paradigm that we have used as our basis for software design and development, but we can also use it with other paradigms too, as long as they result in neat little packages of code that can be used as stand-alone objects, and are not required to be used within the context of a larger system.

Therefore, while code objects are probably directly usable if they are the result of an object-oriented coding approach, only full components will be usable if they are a result of other paradigms. By full components, we mean libraries as well as self-contained controls such as those used by the Visual Basic and Visual C++ programming environments.

It is also important that a specific member of the organization takes responsibility for the management of the object and component archive. This will include procuring tools, establishing guidelines for their use, and training those involved in code creation in the processes required to ensure that the archive is correctly populated with high-quality snippets. This person is also known as the Librarian from Part I of this book.

CREATING AN OBJECT REPOSITORY

An Object Repository should try to satisfy several goals:

- Store source code
- Provide an indexing system for source code retrieval
- Have an integrated backup facility
- Support versioning

Beyond these four basic facilities, there should also be sufficient documentation in the repository's use, along with any conventions that are necessary to satisfy the general working practices of the organization that also have an effect on the efficient use of the Object Repository.

Tools

First, it is always a good idea to use a versioning system, such as CVS (or the emerging SVN), SCCS, or Visual SourceSafe as a mechanism for source code storage, retrieval, and comparison. For those not familiar with the concept of source code control, here follows a short description of what it is supposed to achieve.

In any software project, the code will be in a constant state of flux. Even once the project itself is complete, and the final product is in the marketplace or installed at the client's site, the code that makes up the application will probably continue to change. Errors will be found, and subsequently fixed, new features will be added, and the code base will change.

In a project that is not explicitly keyed toward the consumer market, in which the code that is written is done so entirely on spec for a single client, effective source code control can be seen, in the beginning, as an option, and so no great investment in tools is made. It is often seen as adequate to maintain a list of changes in a spreadsheet, which is then stored along with the project files on a server.

This is not a good approach to change management in a software engineering environment, but is acceptable for managing change for extremely small pieces of larger systems. By acceptable, we mean that there will be cases in which it is necessary to use something, there are not adequate funds or expertise to use another method, and most people will be able to use a spreadsheet. This might not be the case for a specialized piece of software, which may require a certain level of investment to arrive at the same level of functionality.

A spreadsheet works as a way of collecting information, providing a structured way of doing so, along with simple information extraction, sorting, and delivery. As such, it is useful, but it cannot perform tasks beyond those that the users do themselves. Better tools exist that offer more functionality and require less user interaction, and unless there are simply no funds available, it is a good idea to use them from the start.

As always, putting tools to work after the fact, once the information has been input and stored somewhere else, will cost more than implementing the better system in the first place. The danger is that this cost becomes prohibitive, too, and so the move from one system to the other is never performed. At some point, the old system will be proven inadequate, usually at the expense of a contract or project.

Those projects that are designed to create a piece of software seen as the first in a line of such products—such as office automation tools—usually make use of effective tools from the outset. The implication is that, when creating a single product for a client, once the project is finished there is no need to retain the code within a source code control system, unlike evolving products.

However, when the source code is part of a wider Object Repository, where it may be a part of multiple projects, some kind of source code control becomes strongly advisable, particularly within our paradigm of object reuse. In essence, a source code control tool automates a number of tasks, including:

- Change tracking
- Patch generation
- Version management
- Change rollback

Source code control systems all work in roughly the same manner; that is, a set of underlying features describes the technology. How the technology is implemented, and the exact mix of features that are available will change from package to package.

The basic principle of a simple single-user source code control system is that pieces of code are checked in and out of storage space on a centrally managed server. If a piece of code is checked in, it is safe from editing. If it is checked out, then it can be modified, usually by only one person at a time, until such a time that it is checked back in again.

Code that is checked out, therefore, may not be modified by any other party until it has been returned to the control system. The checking in of a piece of modified code is where the magic starts. The source code control system compares the checked-out version with the previously checked-in version, and makes a note of the actual changes that have been made, by whom, and when.

The systems generally allow comments to be added to each check-in operation, so that an accurate history of the entire source code tree is maintained.

Once the principle is understood, it is an easy step to imagine cases where the same piece of code could be checked out to multiple individuals. As long as they are all working toward the same goal, and do not attempt to work on exactly the same lines that their fellow programmers are, it is an easy enough task for the system to merge the code together upon checking back in.

There are also more complex features such as splitting the tree such that new branches of code can be started, and take advantage of changes in other branches as appropriate. Branching and remerging code in this way is very complex, and is best described in the documentation that accompanies the source code control system that the target organization chooses to implement.

Recent tools such as Subversion (SVN) add the ability to have more than one copy of the source code checked out at a particular time, but these rely on the possibility to retain a working copy so that changes can be detected automatically, and updates applied as required.

Allowing this to take place will inevitably lead to cases where two programmers have been working on the same file, and on the same area of that file, such that the source code control system (SVN, for example) can no longer actually decide which version is correct.

At this point, it will usually point to a conflict, and expect the user performing the update (or committing the changes) to choose between two versions of source code. In theory, and especially within the paradigm that we are using in the context of this book, this should never happen.

It usually happens when a programmer finds an error in work done by another programmer, and adjusts it. The original programmer might have spotted the error and fixed it, at which point the source code control system will find two conflicting changes.

What should have happened is that the programmer finding the error should have checked with the original author that the problem has been located and will be fixed. At which point, the original author of the code should fix the error, and check his code back in before the code is changed further.

One other use for the version control part of the source code control package is in creating patch files. A patch file is a list of changes that, when applied to a piece of code (either source or real application executable code) can change it such that it is brought up to date.

One of the reasons for doing this is to avoid having to ship the entire package every time a change is made; just the patch executable needs to be shipped. The executable contains the data that needs to be applied to the existing version to upgrade it, and usually a few routines that verify that the patch is being applied to the correct version of the code, and that it has been correctly applied once the patching routine has finished.

Another reason for using patching is that it reduces the time needed to download a new version of the software, which becomes important in companies using a repository where changed objects could be used in a variety of different applications, each with different user bases. It is much more convenient for the users to download patches than for the company to have to ship new versions of the software.

There are some small points to bear in mind when choosing and using the source code control system, however. First, it will only ever work if there is one source code tree on one server. This introduces a central point of failure. If the server malfunctions for some reason, or the hard drive fails, the results could be catastrophic. Hence the need for a robust backup system based on file copy and archival.

Second, versioning and change management work best when the files are reasonably short; that is, when a single file concentrates on a single subject, be it a piece of text or the source code for an object. The problem is that when files contain many different subjects, the possibility that they will need to be changed in multiple places at the same time increases exponentially. The result is that checking in and out becomes a lengthy procedure since each change needs to be analyzed and documented.

A side effect of this is that the more complex an object implementation becomes, and the more useful it is, the more individuals will want to make use of the code. Subsequently, the likelihood that this individual will want to add to the object, upon finding that it needs to be refined for his particular application, also increases.

This can be the beginning of a cycle in which the code is permanently checked out, and if left to spiral unchecked will lead to deadlock situations in which nobody is able to update the code, even if an error is found. This is detrimental to both the quality of the code and the efficiency of the development cycle, which are both issues that the Object-Oriented paradigm and object archive are supposed to help solve.

Even if an advanced multiuser solution is in place, there is the possibility that multiple conflicting updates can cause problems, but respecting the guidelines within this book should be enough to solve these and any deadlock situations pointed out in the preceding paragraph.

The solution to this problem is to either take advantage of the source code control system's branching capabilities, or encourage object subclassing (inheritance) as laid out in Chapter 9, "The Object-Oriented Paradigm," and Chapter 12, "Coding and Language Choice." If strict subclassing guidelines are adhered to, a single object will have a reasonably small number of new features, thus reducing the chances that the same object will need to be checked out by two individuals simultaneously.

Finally, the overall success of the source code control system will be based on the willingness of staff to participate in the corporate source code control philosophy. This means that each member of staff has to be aware of the consequences of not checking in and out source code in a consistent manner.

It is expected that the specific standards and guidelines relating to source code control are stated as part of the document set discussed in Chapter 2, "Standards and Guidelines." This ensures that both the target organization thinks through the issue properly, and that the staff have an easy-to-follow set of rules by which they should work, and whose performance can be measured against.

Implementing Tools

There is a balance to be struck between automated source code management and source code management that is based on using software application tools augmented by human intervention. The two extremes are easy to appreciate:

- Manual system using a spreadsheet and human librarian
- Fully automated system with little or no human interaction

The issue is that the former is expensive to maintain, but cheap to set up, while the latter is cheap to maintain but relatively expensive to set up. The reality is that, whatever system is put into place will need to be maintained by a human, usually the person who set it up.

Therefore, the practice we favor is a semi-automated tool that is capable of helping the human librarian to manage and control the evolving software repository, but will defer to that human rather than trying to provide a fully automated repository in its own right.

The justification for this is that we are adopting a specific paradigm for the creation of software in which we place emphasis on the procedures and processes that rely reasonably heavily on human interaction through a central point. We are also trying to establish an environment in which it is possible to cross-pollinate multiple projects while at the same time promoting high-quality software development.

For these reasons, we advocate partnership between the librarian and the source code control system that is used to manage the repository. This is distinct from the source code control system that is used to manage the code in individual projects, which can be more automated since the librarian will only be involved with the finished product.

Indeed, it is likely that there will be two separate systems in place—one for the software engineering in the large, and the other for the micromanagement of individual software projects. Essentially, inside each team we are relaxing the control slightly to allow them to concentrate on the programming and engineering rather than the source code control, while ensuring that the librarian can concentrate on the big picture, rather than being called in to help with code management on individual projects.

Directory Structure

Unless the repository system put in place is capable of identifying and linking individual files, a search of the available objects will eventually lead the programmer to a location in which he will expect to find the code and documentation for the object he wants to make use of.

Part of being able to find this information is knowing where to look, and a good directory structure that has been properly identified and documented will make this easier. It also provides a clear standard so that programmers do not have to define their own structures, leading to a much cleaner file system structure.

There is also an advantage when working with source code control systems—versioning and backup. This is particularly the case if a semi-automated or manual backup system is in place. For example, in the case of a semi-automated system, the software will need to be told where the information that needs to be backed up is stored, and in the case of a manual system, the user will have to search through the structure looking for files to back up.

These essential chores are much easier to manage if the basic directory structure is standardized across all objects stored in the repository. For individual objects, four basic directories will be required:

- Source code
- Object code
- Executable code
- Documentation

Depending on the version control philosophy of the organization, this tree might be replicated for different versions of the object, or version-specific directories might instead be introduced underneath each of the preceding subdirectories.

The two possibilities are shown in Figure 11.1, which illustrates how different they can be. Which one is chosen will depend on the versioning practices. In general, if there are going to be many minor versions during the lifespan of the source code, then the structure by version will be more appropriate.

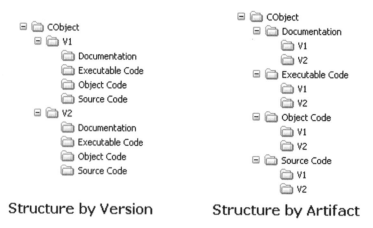

Structure by Version   **Structure by Artifact**

FIGURE 11.1   Different directory structures for the Object Repository.

However, if the emphasis is on the artifact types, and maintaining cross-version congruence, then the structure organized primarily by artifact may prove more useful and easy to implement, navigate, and understand.

Closed Systems and Proprietary Interfaces

It is tempting to look around the marketplace and find a supplier willing to sell an all-in-one solution that can appear to solve all your problems. One probably exists. However, closed binary systems have one devastatingly simple but somehow quite frightening drawback: if the recovery system fails for some reason, there is no way of getting the data back—it might be lost forever.

This is why programs like Visual SourceSafe® only tweak existing files. They do not rush off and encase your entire source code tree in some kind of proprietary file structure; they use what is already there and augment it in an intelligent way.

Visual SourceSafe, for example, manages several aspects of the file attributes to ensure that the source code tree is properly maintained. When a user checks out a file, the copy held by the VSS system is made read-only. At the file system level, this means that the specific attribute that informs the operating system that the file should not be changed (usually called the read-only flag) is set.

Consequently, neither the operating system nor the VSS system is able to change the file. Of course, an advanced user could change the attributes such that the file could be written to as well as read from, which would render the entire process useless. This is another reason why guidelines need to be put in place so that users know how the system is to be used.

When the file in question is checked back in again, two things happen. First, it is compared to the existing version, the changes noted, and the new file copied onto the old one. The read-only flag remains checked, but the VSS system identifies the file as checked in, so that it can be checked back out again by another user as required.

Second, the archive attribute is set—this means that a program scanning the directory tree can identify those files that have changed. This is important for a backup system since it means that a differential backup can be taken because the program can identify files that need to be backed up.

Following this mode of operation, it would seem that the most effective backup tool is one that copies the files one by one to an external device. It is simply an automated bulk file copy tool, and many free/shareware and commercial ones exist.

There might also be a temptation to design your own Object Repository management system, and this approach appeals usually to the technically minded rather than anyone else. If this path is chosen, several consequences need to be borne in mind.

First, to be effective, the Object Repository needs to be in place from the outset, which means that the first project for the company is to create the Object Repository Management software. This may consume valuable programmer hours at a time when the company is at its most vulnerable—at the beginning.

Second, there is the issue of closed binary systems with proprietary file formats and user interfaces. If it all goes horribly wrong, the company will only have itself to blame, and potentially be considerably out of pocket.

It is considerably more efficient to use existing tools, often provided by the operating system, to manage the Object Repository. These might include bulk file copy to external media using the built-in directory copying mechanism of the underlying operating system, coupled with a plain-text file management system for backing up.

If compression is required, judicious use of an industry-standard program such as WinZip can be effective, and at least it has the advantage of being well supported. Otherwise, an effective backup system simply relies on knowing what has been copied, when, and to where. As we mentioned, a plain-text file can hold this information adequately. The only real area where it is vital that some kind of closed system be used for managing the Object Repository is in determining and noting changes to code when it is checked in to the system (overwriting an earlier version of itself), and being able to search all the contents in an efficient manner.

Even here, however, it is important that the actual sources remain unaltered, and that any changes that are identified are stored in a plain-text file so that they can be re-applied to a backed-up version of the source code if that becomes necessary. Following these simple guidelines will ensure that the most vital part of the company's assets—the source code—remains adequately protected and manageable.

MAKING SOURCE CODE SEARCHABLE

Source code is not known for being self-documenting. As such, unlike text, searching for a specific piece of code is unlikely to yield a useful result unless steps are taken to ensure that the code is embedded with pieces of information and text that enable the user to search for meaningful items within the source code files.

Filenames and Comments

The first level at which information can be embedded is in the filenames that are to be used in creating the object that is to form part of the archive. If, for example, we are creating a set of sorting routines, we might place the code in files with meaningful names such as:

- Quicksort.h
- Bubblesort.h
- Sieve.h

These days, when filenames can extend to 256 characters and beyond, there is no excuse for not using filenames that describe both the contents of the file and the project to which it belongs. The organization can find their own particular way of elaborating filenames such that their contents can be easily discerned from the filename alone. However, some part needs to accurately reflect what the source code contained within is supposed to implement.

The same goes for the directory (folder) structure within which the object, which is likely someone else's entire project, is contained. As we discussed previously, it is not advisable to put all the code in a single file. The principles of object-oriented design lead to subclassing objects into new implementations that inherit the capabilities of the parent, while adding new functionality and overriding some existing features. Accurate file naming will provide a first clue as to whether the code is of use in a given situation.

If we look back to our example of a sorting library that is designed to be reused, we will note that we have said nothing about the data that is being sorted in the filenames. Of course, we could create a set of files such as:

- C_String_Quick_Sort_Array.h
- C_Integer_Quick_Sort_Linked_List.h

These names reflect exactly what the routines contained inside them are capable of sorting, but they quickly become cumbersome. Taking the naming convention to such a conclusion is probably not the most efficient way to solve the problem. The more typing one expects the programmer to do, and the more elements there are in the filename, the more likely it is that errors will creep in at the programming level.

They may not be errors that impact the eventual correctness of the code; after all, the compiler will likely catch the error long before a working program can be built. However, typing errors waste expensive programmer hours, so reducing the amount of typing required is always a good course to follow. There is a balance to be found, as in so many areas of software engineering, between information transferal and ease of use.

Filenaming conventions need to allow for the most information to be relayed to the reader, while keeping the filename itself to a reasonable length. Ways to ensure this include specifying a set of accepted abbreviations that can be used in naming files and are well documented and understood by programmers.

We can actually help find relevant areas of source code by inserting plain text in the form of comments within the source code. Rather than creating a long and complex filename, we should rather insert comments that elaborate on the filename.

Again, we have given many indications as to how the organization can approach the problem of determining the exact comment guidelines to impose upon their programmers, but there are several key areas that need to be addressed to make the code easily searchable:

- Input data/type
- Output data/type
- Process
- Result

By way of example, if we want to create a series of bubble sort functions for reuse, which are capable of sorting different kinds of data, we might store them in a single file (say, C_Bubble_Sort.h), and then have a piece of comment preamble before each function that details what each function is designed to do:

```
// Bubble Sort for Integer Arrays
// Input Data : an integer array, the size of array
// Output Data : the integer array, sorted (none)
// Process : sort the array
// Result : true if sorted, false if an error occurs
```

Another use for comments is to determine where the code comes from originally. This is useful because if a fellow programmer happens upon the use of a function that he was not previously aware of, he can then locate the specific object in the repository by following the commented reference in the source code that he is examining.

Variable, Object, and Class Names

This should be obvious—using obscure naming conventions for objects, classes, and variables will cause the source code to be unreadable to the point of obfuscation. However, having an actual convention will mean that searching for specific types of data will be much easier. One such convention, known as Hungarian Notation, is common in Microsoft platform programming circles, and works on the principle of prefixing variable names with specific sequences of lowercase characters, depending on their type:

sz: Null terminated string.

n and l: Integer and long integer.

f and d: Floating-point and double-precision floating-point.

One reason not to use a system of type prefixes is that, in an object-oriented programming model, we no longer need to know what the type of specific objects is, only the various operations that we can perform on them.

There are some exceptions for C and C++ programmers in being able to determine between different kinds of objects:

s: For structure definitions.

c: For class definitions.

This can be retained, as well as the prefix "o". This can be taken to refer to an object of any type, usually initialized using the new keyword. It should usually refer to an instance of either a class or struct object.

Notations like Hungarian Notation are so automatic to some programmers that it will waste many resources if they suddenly have to start using another convention for type and variable naming. It sounds trivial, but these are the kinds of things that are usually overlooked by nonprogrammers that have a profound effect on the efficiency and effectiveness of the programming team.

However, problems can arise when using a notation that tries to tie meaning to the variable names. This is most common when a programmer changes the type of data stored in a variable, but then neglects to change the name of every instance of the variable, leading to two issues:

■ The code may no longer compile (having not changed all the names).
■ The variable name, left as it is, no longer reflects the data.

Due to this, conventions like Hungarian Notation, which try to augment the variable name with additional information, are falling out of favor. Modern programming techniques shy away from this practice in the belief that the variable name should not need to convey information beyond the kind of information it contains.

In the object-oriented world, this is entirely in keeping with the way in which software is being created. In fact, since the only interfaces between two objects are well defined in a compact description of the interfaces, it is unnecessary to provide embellishments to the names.

Having argued against the use of Hungarian Notation, we should also look at some alternatives, or rather, some conventions that make sense in programming terms. Two of them have been mentioned, but are entirely specific to C and C++; that is, the introduction of a prefix that is used to determine a Class or Structure.

There are other conventions, which are sensible and entirely language independent, such as always using descriptive names for variables:

ArrayOfPoints: Rather than ptArray.

While the actual type of a variable may not be important, we should ensure that each variable is named according to its data content. For example, we should make distinctions between numbers, arrays of numbers, characters, and arrays of characters.

However, some variable names are so obvious that it can be argued that a strict naming policy is unnecessary. Consider, for example:

DateOfBirth: Obviously contains a date.

IsEmpty: Reasonably obviously contains TRUE or FALSE.

In the final analysis, we can say that a standard should be adopted, but that the coding style, commenting practices, and language choice may make specific decisions regarding type demarcation obsolete.

Constants

The same rules for choosing a convention for naming constant values apply as for types and variables. Not the detail—Hungarian Notation actually makes very little sense when defining a constant by virtue of the fact that it should be obvious from the name what the data itself represents. A constant named `PI`, for example, is unlikely to be a string.

It is up to the development team to decide what kind of standards they are going to apply when choosing a scheme for naming constants. For example, Windows programmers use a system whereby prefixes are used to identify exactly what the constant is to be used for.

`IDC_`:   A dialog box control.

`IDM_`:   A menu option.

`IDE_`:   An edit box.

`IDB_`:   A button.

At the very least, constants should be in capital letters to set them apart from regular variables.

DOCUMENTING OBJECTS

Besides the actual source code, there should also be a series of supporting documents that detail the design, implementation, limitations, and instructions for compilation and use of the code that resides in the Object Repository. In Figure 11.1, we specified a directory named "Documentation" for this purpose.

What follows is a skeleton format that covers the material that is relevant to the purpose of the Object Repository. Of course, organizations implementing this kind of storage and retrieval model are free to choose their own document formats, but these should give a good starting point.

Programmers' Guide

While it should be obvious from the object's source code, laced with comments as it is, how the object is to be used, there should also be a specific guide to help programmers implement a solution using the object in question.

The Guide can be split into three areas: Information, Usage, and Verification. In the Information Section, we need to be able to specify information that describes the object.

Object Type

This is language dependent, but reusable objects should fall into one of two categories—object code or source code. Anything outside of these (executables, dynamic link libraries) can be considered components, and not objects, and placed in the Component Gallery.

Source Code

This part of the Information Section lists the source code files that make up the object, whether it is a piece of reusable source code (Class, Package, Module, etc.) or compiled object code.

Required Libraries

Following the Source Code, we need to know whether the object has a relation with other libraries before it can be used. For example, if it relies on specific platform-dependent libraries, such as operating system specific facility libraries (network, screen, device), then they should be listed here, even if they are not required in order to compile the source code itself.

Resource Dependencies

The last part of the Information Section should list any nonsource code files, such as graphical images, sound files, text files, and so forth that are required by the object.

Next, in the Usage Section, we are informing the third-party programmer what pieces of code are designed for reuse, what their names are, and how they operate.

For each object (C++ Class, Eiffel Package, Modula-3 Module), we need to have the following:

Exported Functions: Those functions accessible by external code objects.

Exported Class Definitions: Specific objects that can be accessed externally.

Theory of Operation: Typical use cases.

These three sections should give programmers enough information to know whether the code object is fit for the use they have in mind, or whether it can be adapted to fit their specific needs.

Finally, the Verification section lists how the code has been tested, under what conditions, and any exceptions that were found but not repaired. Exceptions that exist and would affect the use to which the object needs to be put should be fixed by the programmer wanting to employ the object. In this way, the entire Repository is in a constant state of improvement.

For each object that we have defined the Exported Functions, Class Definitions and a Theory of Operation, we support it with:

Test Cases: Showing specific usage in action (including boundary cases).

Test Results: Proving that the object is for use.

Open Issues: Showing areas needing development.

The resulting document, comprising the Information, Usage, and Verification Sections, can then be released into the field along with the source code or object library to which it refers. It should be in a format that can be searched by an automated indexing system, as we pointed out in the first part of this chapter, *Creating an Object Repository*.

Design Documents

While we have not explicitly dealt with the design documentation in the skeleton document structures, they should also be placed in the Documentation directory belonging to the object in question. The areas that can be considered for inclusion are:

Requirements Definition: The purpose of the Requirements Definition here is similar to that which we introduced in some detail in Chapter 6, "Requirements Definition," namely to inform the reader of the purpose behind creating the object.

Functional Specifications: Again, we already looked at Functional Specifications in some detail in Chapter 8, "Functional Specification," and this reduced version exists simply to state what the object can do, which solves the problems presented in the reduced Requirements Definition.

Augmented Design: The Augmented Design is a very loose description of how the facilities described in the Functional Specification, which satisfy the Requirements Definition, have been implemented. It should include pseudocode in cases where an advanced proprietary algorithm has been created to solve the problems presented.

Certain aspects need not be described, where they fall within the general science (e.g., linked list, stacks, file storage, etc.), except where they deviate significantly from the current state of the art.

The idea is that these informal design documents represent a work in progress—that is, the design might have to be changed depending on the future evolution of the object with regard to new functionality to be added or errors to be corrected.

However, there will come a time when the current version of the object should be left in a static state, and a new object be created, which might depend on the existing object, to service any new functions that come to light. For example, an object that represents a printing device might be extended to cover plotting devices as well. Equally plausible is a scenario that involves creating an object that derives some of its functionality from the original printer object, but specifically encapsulates the functional description of a plotter.

A natural extension of this decision process is to separate objects when they become too unwieldy. It is a policy decision that of the two paths to take, depending on individual circumstances, it may be easier to restrict the functional areas of objects from the outset rather than try to carve them up at an arbitrary point.

Therefore, to bubble sort a list of integer values, we might come up with the following outline design documents:

Requirements Definition
    To sort a list of integer values.
Functional Specification
    Supported functions:
        Create a list
        Add items to the list
        Sort the list
        Retrieve items from the list

Supported data types:

Integers

The list should be able to contain a maximum of 100 integers.

Augmented Design

The data shall be stored in an array, 100 elements in size.

Sorting mechanism: bubble sort.

Algorithm:

```
while ( no more swaps needed )
      if element_n > element_{n-1} swap element_n, element_{n-1}
```

There is no need to describe the mechanisms for inserting and removing data from the array, since that is part of the standard science associated with programming in general.

THE COMPONENT GALLERY

The principle difference between the Object Repository (OR) and the Component Gallery (CG) is that the OR allows programmers to make use of source code, while the CG merely contains a set of libraries that are only available in object code format.

These might be commercially obtained; a number of companies specialize in the production of components for a variety of applications that can be purchased and plugged into applications. Some basic components might include graphing, Internet facilities, reporting, and encryption.

Directory Structure

We can follow a similar directory layout as that which we saw in Figure 11.1, but with the source and object code trees removed. As can be seen in Figure 11.2, we are left with a slightly simpler layout to follow.

This being the case, the librarian might decide that it is not necessary to follow the same layout as was chosen for the Object Repository. For example, if the Structure by Version had previously been chosen for the Object Repository, then the librarian might opt for the Structure by Artifact for the Component Gallery to keep artifact types together.

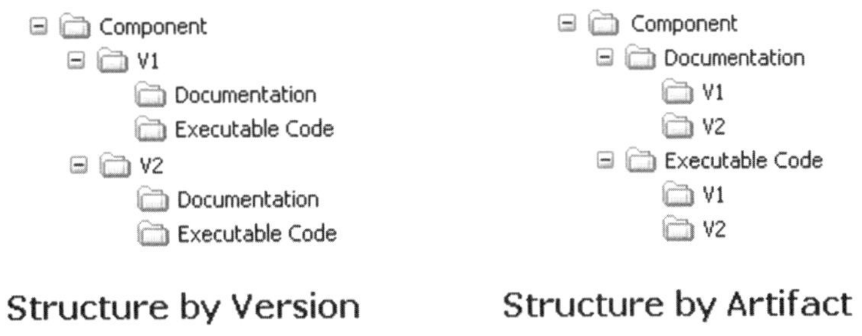

```
⊟ 🗁 Component                    ⊟ 🗁 Component
   ⊟ 🗁 V1                           ⊟ 🗁 Documentation
      🗁 Documentation                    🗁 V1
      🗁 Executable Code                  🗁 V2
   ⊟ 🗁 V2                           ⊟ 🗁 Executable Code
      🗁 Documentation                    🗁 V1
      🗁 Executable Code                  🗁 V2
```

Structure by Version   **Structure by Artifact**

FIGURE 11.2 Different directory structures for the Component Gallery.

Documentation

In a similar way that we defined documentation for storing the Object Repository, we also should define documentation for the Component Gallery so that individual components are easy to find, and their purpose and use well defined.

Object Type

This should state whether the object is an executable, a dynamically linkable library (DLL) (Dynamic Link Library in the Microsoft Windows operating system), or some other form of entity, such as a plug-in or applet. The key distinctions are that an executable is often a stand-alone tool built with a specific purpose in mind, whereas a DLL is an operating system-specific extension to applications containing code that can be called from any application managed by the operating system.

Plug-ins and applets require specific support within the application, such as a set of functions, usually provided by the component manufacturer, designed to be included in a project that needs to read and execute the plug-in or applet. These are usually application specific, although the widespread adoption of Java has meant that some operating systems are capable of executing Java applets as if they were native applications.

Required Libraries

This part of the document needs to detail any specific third-party libraries that need to be present in the target system using the application, or the system used to build the application. These might include operating system-specific functions, such as those parts of the operating system deemed optional.

Theory of Operation

This part of the document gives an overview of the problem that the component is designed to solve, or the way in which the tool is to be used to achieve the desired results.

It can be either aimed at the end user, being the developer wanting to use the tool, or a set of guidelines designed for a programmer to read that dictates the way in which the component is to be integrated into a larger system.

Command-Line Options

In the case of command-line tools that are stand-alone executables, there may be a part of the document designed to inform the end user about the use of any specific command-line options designed to change the way in which the tool operates.

This section is optional.

Test Cases

Independent of any tests carried out by the original developer of the component, there should also be a collection of test case descriptions and results carried out by the users of the component.

These tests are necessary because they will be performed at the same level of rigor as the rest of the system that is being developed, and the objects that are being implemented in order to create that system. Thus, by including the component in the gallery, the other members of the organization can be sure that it has been subjected to the same level of quality assurance as any other pieces of code they might choose to use.

It is also useful in determining any limits on the operation of the tool, or use of the component that the original developer had not envisaged simply by virtue of the fact that every user will require the system to operate in a slightly different fashion.

Most of these we covered previously in the Object Repository part of this chapter, but the slightly different usage leads to some notes to make. First, only objects that are directly useable should exist in the Component Gallery, which usually means executables (tools) or libraries (and resources).

Second, in the case where the object is an executable, the user needs to be informed as to how the command is invoked, what parameters can be fed to it on the command line, and whether it returns a value upon completion.

It would also be a good idea, in the Theory of Operation, to note any run-time effects that the executable might have on the operating environment, such as memory or disk usage, and a list of error conditions, how they are arrived at, and how to test for them.

Finally, independent tests must be carried out that validate the fitness for use of the component, exactly as if it were something created internally. After all, it will probably find its way into the production cycle of other projects that will rely on the component behaving in a correct manner.

In the case of purchased components, a single document, provided by the developer, might cover all the appropriate topics, but might not follow the guidelines set out previously. It then becomes a policy decision as to whether the information needs to be extracted and inserted in documents that follow the same skeleton format as components developed internally.

Searchable Executable Code

There is also a mechanism by which programmers can render their source code more easily searchable, such as being careful to choose variable and function names that accurately match the purpose of the code snippets to which they refer.

The usual rules for variable naming should apply to those pieces of code that are supposed to be searchable. Comments, naturally, will probably be compiled out of the final fragment, so it is really up to the programmers to ensure that those pieces of information that might help a plain-text indexing system are embedded in the actual source code.

Some tricks include defining constant strings that contain comment-like statements, which will be compiled and placed in the resulting code as plain-text strings. In fact, hackers often use this kind of searching mechanism to try to locate pieces of code that offer security features such as checksums and time-limiting execution.

It is, therefore, a technique that cuts both ways. By making the code searchable, one might also be making it prone to being hacked, since the final version will also contain those pieces of code that appear in the components being reused.

As a final note, it should be pointed out that this is very much a last resort, and only to be used if the source code is not intended to be made available, or is not available. The best way to make sure that a component is findable is to derive a description from the source code and add that to the repository of searchable documentation.

Even in cases where a closed binary library is being used, there will be (have to be) a definition of how the library can be interfaced with, and it is this that can be used to derive a description that can provide a searchable version of the component in question.

SUMMARY

This chapter covered effective storage, backup, and management of components that might be used in software projects. We listed the types of supporting documentation, ways in which the directories might be laid out, and provided some pointers as to how the entire tree should be backed up.

System managers, of both mainframe and desktop machines, will have their own ideas about how to copy entire directory trees of information onto archival material. From the Unix tar utility to third-party solutions such as PKZip, there are many different ways to achieve the same result.

We said very little, however, about how the visual front end should be achieved. This will vary with the kind of system that is available at the time of implementing the first archival solution. It is recommended that the entire archive have, as a minimum, a network of HTML (Web) pages that can be used to guide the user.

Therefore, for each area of the archive there should be an index page that provides a link into the actual directory where the resource is located. There are a few things to bear in mind:

- Links should always point to the most recent version.
- Some files will not display correctly in a Web browser.

The clear advantages of using such a front end are that the resulting network of Web pages is searchable using standard tools (such as those from search engine manufacturers such as Google), and they can be edited using standard tools such as Microsoft Word.

There are even some specific CGI scripts that can be used to provide a user-friendly change management system (such as that on display at SourceForge.net) that includes provision for releasing updates, searching through information, and archival and source code indexing.

One other aspect that we neglected is the effect that the object and component archive has on quality assurance. Many projects have encountered problems because some form of synchronization issues had been introduced between different binary versions.

For this reason, it is usually beneficial to always begin a test cycle from a virgin build, created in an environment other than that used by the development team. Within the confines of the object and component archive paradigm, this may not be possible, unless all the components used are Open Source.

In addition, if fully tested components and objects are being used, it is unnecessary and inefficient to perform a complete build. After all, they have been reused from a point of view that they are perfect, or at least that their defects are well known and publicized through the object and component archive system.

12 Coding and Language Choice

In This Chapter

- Introduction
- Language Layers
- Specific Languages
- Choosing the Glue
- Comparison of Modern Languages
- Summary

INTRODUCTION

Before actually starting any programming, it is important to establish what the mix of programming languages used is going to be. Some projects, if they are merely gluing together sets of previously written objects and components, can be developed entirely using a high-level language that may not necessarily be compiled—it could even be an interpreted language, either Web-based, such as PHP or ASP, or one of a variety of interpreted client side languages such as Tcl/Tk.

This chapter describes the different layers of a given system and how these layers will reflect in the programming language used to develop the system. While systems can be built using a single language, in many cases it will be beneficial to select languages that are most appropriate, for performance reasons in highly graphical systems, for example.

The language need not even be one that already exists. Some applications need to be extendible by the client to reflect their changing operating environment. Financial messaging systems, for example, tend to be reasonably fluid, sometimes changing as much as twice per year. In such systems, being able to add features to the protocol used for messaging is often scripted with a protocol definition language that is peculiar to the system in place.

This means that the developer does not necessarily need to be involved for many of the changes in the system that are prescribed by a change in the messaging protocol. Other applications, such as those in the graphical processing industry, also use scripted plug-ins for special effects either on moving or static images.

If a variety of languages is to be used, it is also probable that a single language will be chosen as the glue to stick all the various other pieces together to create the final product. Choosing this glue language is very important, since it must be flexible, easy to write, and offer enough power to combine the other components.

It is assumed that the actual hard logic and heavy processing of the system is not performed in the glue language, but in specific components that have been designed and implemented to solve specific problems within the final system.

LANGUAGE LAYERS

There are several layers of computer language, and the key to understanding how software engineering and programming work together is to look at the entire concept as if it were an onion. In the center of the onion, we have the central processing unit of the computer, surrounded by whatever hardware it needs internally that will enable it to successfully perform the tasks for which it has been designed.

Around that core, we have a set of instructions that are understood by the computer, but not by anyone else. In fact, some engineers can write code that is directly executable by the computer, but this is increasingly rare as machines become more and more sophisticated and the number of instructions that they understand expands.

From this machine code, out to the edge of the onion, there is a process of abstraction, which means that the languages become less and less related to the machine and more human oriented, on top of which are layers that translate from the human oriented to the machine oriented (such as Java) in an attempt to remain platform independent.

This is necessary because not all machines speak the same language, so we either need to convert our high-level language into the flavor that they are expecting, so that it can be executed (compiling), or we need to write and compile an application that can translate, at run time, the language into actions on behalf of the processor (interpreting).

Clearly, there will be performance implications for using machine languages, compiled languages, or scripted languages, and this section of the chapter will attempt to divide all the possibilities into three distinct areas.

Machine Languages

The most efficient programs can usually be created in machine code. By efficient we usually mean fastest to run. The reason for this is that humans are actually quite good at refining code such that it is arranged in the most effective way for a machine to translate into actions.

One of the best ways to achieve a trade-off between the advantages in execution speed and development time is by compiling the code first into something that the machine can understand and then refining it so that it is even more efficient. Except in very specialist cases, compilers are actually getting consistent and effective enough that this is usually not necessary.

The areas in which an advantage can usually be gained are in driver and high-end graphics development, or areas in which speed is vital. There may also be cases where no compiler exists for the hardware because it is proprietary or not widely used. In these cases, it can be more efficient just to write a compiler.

Writing in machine code is very inefficient in terms of coding time. The languages tend to be very complex, and built up of many different specific keywords and concepts that are not easily understood. Above all, it simply takes a lot of time to write code this way, and modern computer languages are far more efficient.

Assembly Language

Directly above pure machine code is assembly language, and it is this that is most commonly used by those who need to be able to control the executable code that is used to create the application. Assembly language differs from machine code in one respect—it can be edited as words on the screen, while machine code has to be edited at the byte level.

The assembler performs two tasks—it takes the assembly language and converts it into the specific byte code that the processor understands, and it prepares an executable file that follows the specific format of the underlying operating system.

In most, although not all, programming paradigms, the operating system runs the executable. Some of the code may be directly executable by the processor, for the sake of efficiency, but the operating system actually prepares the memory, loads the software, and manages the allocation of resources, before helping it when it decides to terminate.

Generally, this task is performed by the linker in the last step in the compilation cycle. The input to the process is the assembly language (or object code) for all elements of the system, including any libraries and subsidiary modules that contain code that is needed for the system to run.

Operating System Concerns

There are as many different formats for the executable file as there are operating systems, which means that writing code in assembly language is very system specific. In fact, it is fair to say that code written in assembly language will be almost impossible to port across platforms, even if they represent two operating systems from the same family.

This means that they are not appropriate for use in an object-sharing software engineering paradigm. There is less and less need, except for specialist applications, to use assembly language, but in some environments, such as embedded systems, video game and simulation, and hard real-time systems, it becomes inevitable.

There is, however, one final advantage: by using machine code, it is not actually necessary to have an operating system at all. In fact, if the program is capable of managing its own resource allocation, and does not need to rely on any specific services that might be offered by the operating system, then it may be able to run on its own, which can be an advantage for certain projects.

Compiled Languages

The next level of abstraction away from the machine is in using a compiled language—typically, these are plain-text editable files or possibly some form of fourth-generation language that is based on diagrams. They are hence much easier to understand than assembly language that makes writing code more efficient.

However, depending on the compiler, the language construction may not lead to particularly efficient code. It is fair to say that the more abstract the language becomes, and the more innovative it is, the less efficient will be the compilation. Thus, a new language based on a very abstract concept using near natural language ideas and terms will compile to less efficient code than a language such as C, which is close to its origins and is supported by tools that have been in existence for about 30 years.

Most projects will use compiled languages as the core. They are easy to port, as long as the toolset that supports them supports the target environment, and provide an interface between the developer and the system that is easy to absorb, while still retaining concepts that make them efficient to compile.

The nature of the support and organization of compiled languages also means that they are perfect for use in code-sharing initiatives, such as the Object Repository concept that we introduce in this book as one of the pillars of software engineering on a corporate level.

Byte Code and Virtual Machines

Some compiled languages, such as Java, do not compile to native code. That is to say, they are reduced to code that will not run on the platform as it is, but must be interpreted by a special kind of processing application called a *virtual machine* (VM). The key principle behind this approach is to allow the same executable to run on any platform for which there is a VM available.

Some people might put these languages in the same basket as scripted or interpreted languages; others will place them in the compiled languages basket. In principle they lie somewhere in between. They are, however, particularly suited to the object-sharing ideology, since as long as the operating system contains the correct VM, they can be used, without even being recompiled, over and over again.

Interpreted Languages

The last type of language that we will consider is a family known as interpreted languages, which includes all those that need a run-time portion in order to execute them, such as application enabling languages, and those that are not compiled but attached to a front end that interprets them.

These are different from approaches such as Java that use a VM, because the compiled version of Java is a more efficient rendition than the language was originally. This may not be the case for interpreted languages that suffer from the additional problem that the application that launches the program may also not be efficiently written.

In the worst case, it could also be a byte code compiled application, requiring a VM, in which case the code has to be interpreted twice before it is actually turned into an effect or action.

Interpreted languages in wide use include JavaScript, and many Web languages such as Perl, Active Server Pages, and PHP. They provide an easy way to develop very powerful cross-platform solutions, which is, of course, the goal of using an interpreted language.

Security and Performance

There are two big disadvantages with using interpreted languages, and it depends on the exact way in which they are used as to whether either presents a problem.

The first is that because they are essentially plain-text files, they are not secured in any way. It is true that it is possible to reverse engineer a piece of compiled code to get back to something that resembles a programming language, but it is unlikely to yield anything usable due to the vagaries of the compilation process itself.

However, when an interpreted language is used, the code is available for all to see if it is, for example, embedded in a Web page or handed over as part of the deliverables in a scriptable system. If the script is running on a Web server, with only the end result being handed to the browser client, then only a concerted attempt to download it will actually give access to it, and it is unlikely that this will prove possible.

Performance is an issue that will affect every platform, but to what extent it does, may not matter. For example, it is unlikely that Internet users will notice a difference between the reaction time of a scripted and nonscripted environment. They will experience a bit of lag on the line anyway, which will mask the execution speed differences.

However, on a local system, the user may well be aware of a difference in reaction speed between two systems, and if the environment requires high performance, there is a distinct possibility that an interpreted language will not provide enough performance to satisfy.

Scripting Engines

To get around these two issues, and in an attempt to retain the ease of use of an interpreted high-level language over a traditional programming language, the developer can choose to use a commercial scripting engine. This is somewhat like a byte code interpreter, except that no compilation takes place, except just before the script is intended to be used.

Internal Compiler

Inside the developed and distributed application is a compiler that turns the high-level scripting language into executable actions, and then executes these actions—it is a halfway point between a strict byte code interpreter and a real programming language.

There is also the possibility that, rather than containing a compiler inside the application itself, the languages are turned into byte code that is then distributed by the company as part of the application. Hence, they will be able to extend the functionality of the system without changing the core source code, but the client will not be able to have access to these customizations.

Therefore, the use of a scripting engine provides extensibility, security, and better performance than an interpreted language. It can also mean that a programmer is not required to produce a customized version of the application for other clients, which is something that fits nicely with the reuse model and traditional commercial frameworks.

SPECIFIC LANGUAGES

Of course, in any system there will be the possibility that it needs to make use of some specific languages that deal with issues such as document creation, data manipulation, and so forth. In fact, most of the communication to external services (such as databases) will rely on a language that needs to be integrated with the end system.

This can be done on an ad-hoc basis, at run time, or designed into the system and compiled as if it was any other programming language. The only requirement is that the manufacturer of the third-party tool or service gives programmers some way to interact with the application via a connection to it.

There is also a variety of languages that are concerned with the presentation and communication of information. Consider the file format that provides data exchange between applications such as word processors—this can be considered a language of its own, albeit proprietary.

Some languages, such as PostScript, are designed to aid in communication of the document to printers, and these are languages that began as proprietary creations that have moved into the mainstream. There are also some libraries available that can be used to display PostScript documents.

Therefore, while the specific languages themselves are probably not appropriate for use in the Object Repository, it is probably worth building up some areas of the Component Gallery to contain libraries that can handle some of the more common types of specific language, which we have split into three groups:

- Document Definition Languages
- Data Management Systems
- Communication Languages

In addition, there is also a whole host of specific protocols that are not languages, but dialogs that enable certain technologies. For example, TCP/IP is a protocol designed to be used in Internet communications, but is not rich enough to be considered a language in its own right. It does, however, provide a sound base for the transmission of HTTP requests, which could be considered a very basic language, and on top of that HTML, which is used to define Web pages and is certainly worthy of the title language.

Document Definition Languages

The purpose of document definition languages is to provide a portable way to display information that can be reused across multiple platforms and devices, usually giving an effect that is constant across those devices.

The definition language will be reusable, as a definition of a way in which the data is to be represented, but it is unlikely that the scripts themselves will be relevant in different situations. There will be, of course, exceptions to this rule, such as where legal pro forma or accounting packages can reuse text and layout, but these must always be augmented with user data.

It is more likely that the library that is capable of producing the actual end result, using the definition language as a kind of instruction set, will be the portion of the system that is reused, and as such, it will form part of the Component Gallery, or the Object Repository, if the source code is available.

Examples of Document Definition Languages include HTML (for hypertext Web documents), PDF (the standard for portable document exchange originally created by Adobe), and PostScript (for printing and raster display).

Data Management Systems

Data Management Systems have an interface language, usually a variant of the popular Standard Query Language (SQL). The purpose of this language is to manage the data that is contained inside the database system. This includes aspects such as table creation, data insertion and removal, and basic retrieval tasks.

Unlike Document Definition Languages, then, they become based on a dialog between two systems—there is a request, and usually a response. The Document Definition Languages, however, are entirely one way—the system merely uses them to create a document that is then validated by the users when they try to read it on another system or using another application.

The difference is that the successful implementation of the interface language can be directly assessed without the need to look to a third application. The language is again a template, a way in which the programmer can seek to define the queries that the database system is to execute, but they contain more than just a write-only document, and will actually cause actions to take place that have potentially far-reaching consequences.

Here again, it is unlikely, except in rare circumstances, that the scripts themselves will be reused. They will be created by the application during run time, and will only have relevance to that particular application.

Therefore, the library provides the interface to the external database system that is inherently reusable, and that will, again, be a part of the Object Repository or Component Gallery.

Communication Languages

Finally, we have the set of languages that we have chosen to call Communication Languages, and this is the most varied group. On the one hand, there are the raw communication enabling protocols that allow a dialog between the requester and the server, such as HTTP.

HTTP allows the Web browser to request various bits of information about an artifact, as well as setting up a dialog that ends in the requester receiving that artifact (or a copy of it) via a network connection. Such protocol implementations are usually accessible via a standard library that is available with most development environments.

Indeed, it is possible that the development language being used will attempt to shield the programmer from needing to understand the underlying HTTP protocol itself, by encapsulating the behavior in an object that is capable of providing a buffer between the technical side and the logical side. This object forms the reusable part of the system, or language, not the protocol.

Then there are languages that are designed to convey information in a structured manner. The leading light in this field is a language known as XML (eXtensible Markup Language) that contains in its definition the possibility for the user to extend the definition of the language. Thus, it is possible for the application to extend the way in which the language is used by defining new pieces using a mechanism that the language defines.

This makes it a very powerful tool when information needs to be transmitted in a way that should not necessarily be tied to a specific collection of data. As such, it will usually be the library that facilitates the parsing and use of XML that is reused, rather than any XML "documents."

However, unlike some of the other specific languages that we have seen, it is possible that the XML documents will be able to be reused within an application; that is, although the definition remains static, so does the use, it is only the data that changes, and so a company can, in a sense, adopt a particular XML document to ease the transmission of information between applications.

CHOOSING THE GLUE

Once the most appropriate objects have been chosen, in some cases providing support for scripting or definition languages and in others simply offering a service that is required by the system, then a language has to be chosen that will tie them all together.

It is highly unlikely that the objects will be capable of conversing with each other in a way that is conducive to achieving the desired result without some kind of enabling technology that holds them together. This enabling technology will also probably need to contain a minimum of logic as well, either to process or cause the processing of the input and output data exchanged between objects.

We call this the glue code, since its purpose is to hold the system together in a coherent manner. The more glue code that is used, and the less "real" code, the better. The reason is that we assume that the Object Reuse paradigm has led to a rich collection of well-tried and tested stable objects, which can all be connected together via glue code that serves no purpose other than to link the pieces of common logic together and drive information from one to the other.

In reality, of course, this will be more complicated, and it is likely that at least some of the glue code will provide some form of service that is so specific that to create an object from it would not be appropriate.

Compiled Glue

The most common way to glue the objects together is by using a compiled language, and for this to work, several criteria need to be satisfied, depending on the kinds of objects and components being integrated.

For example, if the system consists of a number of objects for which the source code is available, then it may be a simple matter of including the source files in the project and compiling the whole collection together as one big application. This is the most likely approach in cases where the reused components are individual functions or complete objects, such as Java or C++ classes.

It may be preferable to compile the objects separately, however, and export a suitable interface that can be accessed at compile time from the glue code, and the object code linked into the project at a later stage in the compile process. This helps to keep the objects isolated from the main code base.

This option is most appropriate if the Reuse paradigm extends to libraries or components for which the source code is not immediately available. This also implies that a well-documented API 5 application programming interface is available, as well as a suitable method for linking, either dynamically at run time or statically at compile time.

One example of glue code is in examining a regular Windows application written in the C programming language. Microsoft provides many different libraries for manipulating the Windows user interface, and they can be statically or dynamically linked into via an API that is available in a variety of language flavors, thanks to the development community at large.

The amount of glue code required to write a Windows application that does nothing is around 100 lines. Compared with a command-line interface program, this seems like a lot, but one of the advantages of the Windows programming paradigm is that there are many features built in to the interface.

For example, to create a full text editor, with file handling, would not require much more than 150 lines of code. The actual edit window, as in the Notepad application, is provided by Windows itself, as a single control, requiring a few lines of code to implement. Clearly, this is vastly different from the command-line equivalent, and serves as a good example of code reuse.

Scripted Glue

In a programming environment such as the one used to control the back end of Web applications, it is more likely that the glue will be scripted. This essentially means that the individual objects that are used to provide services are executable, or completely self-contained scripts, and the glue merely passes information from one piece to another.

This requires that the executables are command-line entities, and that they can be communicated with before, during, and after execution. Of course, this does not necessarily hold for all scripted or executable objects held together in this way, but if interaction is required, then it needs to be catered for.

Again using the Web application development environment as an example, a specific language (such as HTML) may be needed to be used as glue code. The project may consist of nothing more than implementing a Web-enabled feature such as a bulletin board application, with an HTML front end. In this case, the only programming that is required is in manipulating the bulletin board with a set of HTML pages.

COMPARISON OF MODERN LANGUAGES

When deciding which language is appropriate for each project, it is necessary to ask a few important questions to narrow the field. We need to arrive at a solution in which the language chosen is the most appropriate for the tasks at hand—some languages offer advantages in terms of tool support, and others offer advantages in that they can create applications rapidly.

Expertise

Key in choosing a language is considering what market expertise is available to exploit a given language. It is very difficult for programmers who have been trained in one language, and have become proficient at it, to drastically change their outlook to program in a different one.

If the two languages are the same in terms of programming paradigm, it will be easier for a programmer to switch allegiances in this way, but it is still wise, at least in the beginning, to use a programming language that is well established in the marketplace and for which adequate staff will be available.

There will, of course, be cases where the application domain is so specialized that a specific set of languages, scripting languages, or tools are vital to the development of the projects in question; in which case, there may not be expertise readily available, so it will be necessary to train new recruits.

Should this be the case, there is another debate—whether to employ recent talented graduates or older hands who have more experience. This debate will probably come down to a compromise solution where a mixture of experienced and inexperienced staff will be chosen in an attempt to weigh the cost of experience against the gamble of providing internal training.

Support

Similar to selecting a language that has professionals ready to exploit it, the language also needs to enjoy enough support that there are tools, components, source code, and APIs available for it. If, for example, the company is going to be developing video games for a specific games console, they need to select a language that enjoys support for the creation of software for that console.

Support for tools includes compilers or interpreters, resource editors, sufficiently advanced editing and code management tools and environments (at the programmer's workbench level), and any specific editors that will be required for the target environment. For example, resource editors for Windows applications need to be accompanied by resource compilers that can bind the resulting file to the application so that the resources (bitmaps, dialog box definitions, menus, etc.) are available to the application at run time.

Early adopters of a given language will find that they become instrumental in providing the toolset for the rest of the community, and will probably enjoy a lot of support from that community. This may mean that it becomes a more investment-effective way to run a software development initiative, since many of the tools and utilities will be available for a more reasonable price, or even for free.

However, this might turn out to be a false economy, since experience will not be available in the general market nor will the tools be stable, which is why, at the time of writing, most software development organizations use one of a reasonably small collection of well-established languages for development purposes.

Frameworks and Environments

Choosing the language will also be based on the availability of application development frameworks that match the target platform. This is in an attempt to remove much of the more mundane programming chores away from the programmer and into libraries that support features that are common across applications within that domain.

This includes, for example, frameworks such as MFC, that originally started out as a way in which Microsoft provided a set of rich classes for containing applications, and offered many standard features such as printing, a multiple document interface, and standard dialog boxes.

More specific application domains will likely have their own solutions to these kinds of problems, such as development efforts for specific hardware that should take advantage of the possibility to reuse code that is provided by the manufacturer, rather than constantly trying to reinvent the wheel.

In other words, it is of no use to try to take advantage of the cutting-edge facilities provided by a new fourth-generation language if that language is not supported within the target application domain. Unless the features of the language are vital to the application under development, it is best to use well-established languages and environments that are well understood in conjunction with the target platform and problem domain.

Portability

If the resulting code needs to be portable across multiple platforms, the language choice becomes slightly clearer. If, for example, the application domain can rely on the presence of a Java Virtual Machine (JVM) on every platform that it is supposed to run on, then clearly the language of choice will be Java, compiling to byte code.

By a similar token, a decision can be made to use an interpreted language that can be run on any platform for which an interpreter exists. This might be a commercially available interpreter, or one that comes with the development environment and contains a redistributable portion that can be given to clients.

However, seemingly portable solutions are occasionally not. For example, it is well known that operating systems are not binary compatible. Even though a Windows environment might look a little like an Apple MacOS environment, programs written for one cannot run on the other unless they are redeveloped.

This means that you need to consider portability at the application binary level (compile once, run many), and need to be aware that some development environments are not even source compatible (compile many). This is because the programming paradigm for each platform is very different, and the way in which the application interacts with the user and the operating system changes for every platform—from Windows to MacOS, Linux to XBOX, they are all different and share almost nothing in the way they must be programmed for.

Performance

The final deciding factor that will help in selecting the language is performance. By this, we mean performance in the run-time environment and performance in terms of the efficiency of development. A language such as Visual Basic is very easy to use, and highly efficient in terms of developing applications in a way that is logical, well supported, and requires very little direct development effort.

However, the resulting applications have a tendency to perform less well than those written with other languages such as Java or C++. Part of the problem is that the compiler model from VB to executable code means that the result is not very efficient, and part is due to the fact that it is a proprietary language and therefore cannot take advantage of wider analysis that might lead to improvements.

This will also affect the decision to use a scripted or interpreted solution in cases where performance is a deciding factor, and one that might not be solved by simply forcing the program to run on a more powerful variant of the target platform.

The result will be a trade-off between available experience, performance, and efficiency, providing that the decision is not effectively taken out of the hands of the developer due to other constraints imposed by the application problem domain or the client.

SUMMARY

In this chapter, we tried to touch on the issues surrounding how a development language, or mix of languages, can be chosen to best serve the project team. In doing so, we also explored many of the different paradigms that are available to choose from, from the point of view of a project manager. Programmers will realize that the view offered is somewhat generalized, and will have a better insight into the problems facing them.

In the end, the technical programming team will have to have the final word, except in cases where the client has expressed a clear preference for a certain solution that should be adopted. The programmers have to work to realize the project, and they should be able to choose the tools that are most appropriate for the job.

This is not to say that there is no merit in selecting or just evaluating emerging technologies, since the whole field is moving at a such a fast pace that there will always be a new solution just around the corner that can be exploited. However, care needs to be taken when doing so since it will increase the risk associated with the project, possibly to a level that is not acceptable or compatible with the end goal of customer satisfaction.

13 The First Prototype

In This Chapter

- Introduction
- Designing the Prototype
- The Prototype as a Skeleton
- Prototype Layers
- The Demonstration
- Assimilating Client Feedback
- Recording
- Summary

INTRODUCTION

A prototype may the first time that the client is invited to look at something more substantial than a paper-based approximation as to what the final product will look like. As such, it is vital both as a first impression of the usability of the system and providing clues as to whether the developer actually understood what functionality the client wanted, and whether the client has accurately expressed what problem they needed to solve.

This chapter describes how the prototype should be designed, since it is as much a part of the final delivery as the actual product, and how to choose the kind of prototype to produce. This can be as simple as a self-contained product that is built using a rapid application development tool but lacks the complex internal functionality that will be present in the final product.

This prototype could also serve as a skeleton into which the code that actually provides the required functionality can be placed.

The other end of the spectrum is a prototype that is a full implementation of the final product, but is developed using tools that result in a package that does not exhibit the performance required in the production environment.

Part of the reason behind creating a prototype is to offer the client the possibility to alter pieces of the application before it is finalized—either because their requirements have changed or because the developer has misunderstood the purpose of some or all of the application. It also gives both parties a convenient way to measure progress, especially if an approach is taken whereby the prototype evolves from an initial skeleton to the final product, or is a full-featured application in its own right.

DESIGNING THE PROTOTYPE

The prototype needs to be designed in the same way as any other software component, always ensuring that guidelines are followed regarding the documentation and design processes. There is a balance to be struck between having a demonstration model up and running as quickly as possible, yet implementing enough functionality that the client can be confident that the project is proceeding in the right direction.

Design Criteria

Before the design can be created, it is necessary to know what the client expects from it—are they looking for proof of concept, or are they more concerned that the user interface makes the software easy to use?

The criteria that need to be examined are:

- Nonfunctional constraints (performance, platform, etc.)
- Timescale (does the client expect fast results)
- Simulation of external interfaces (other systems, databases, etc.)

Essentially, the design needs to be able to address the functionality of the prototype at a level that the client recognizes as being representative of the potential of the end product. To this end, they need to be aware that the closer the prototype gets to being a working application, the more resources the creation of the prototype will consume.

There are two reasons for designing prototypes that are full systems that show all the expected features of the final product. The first is in cases where the entire system is so complex that there is a real advantage to be gained from producing a quick, low-performance version that proves that the developer is capable of delivering the core functionality of the system.

The second reason is when creating a system that has real-time dependencies that affect the safety of the operation of the system—in particular, those systems that are designed to carry people, or whose malfunction will lead to the injury, death, or environmental damage. A simulated system can be created as a prototype that helps to ensure that the design and operating assumptions were correct and that the system has been correctly thought out.

In both cases, all the externally influencing devices or operating conditions will need to be simulated. This can lead to a deadlock situation in which it is necessary to design, prove, and build simulations of the externally influencing factors to the same high quality as the final product.

In a sense, it is possible to go too far, and spend more time proving that the original design, as implemented in the prototype (simulation), is correct than actually developing the final system. The time may not be wasted, however, and it will certainly speed up the development of the final product if all the design and implementation issues have been solved in the process of creating the prototype.

Some situations may call for a third option, in which the user interface and nonoperational functionality—such as data entry, menu systems, and warning, error, and status messages—is created as a prototype, and the actual functionality simulated in a separate project before combining the two to create the final product.

It will be a collaborative effort to design the approach that should be taken to create the prototype, based on input from the experts in their respective fields. The clients will offer their experience and financial constraints, while the developers will offer their knowledge of the construction of computer software.

Between the two parties, they need to reach a conclusion as to what the associated risks are for each of the three prototyping approaches presented previously, if any, and any practical concerns regarding scheduling and finance that either side might have.

Documentation

It is important that the documentation created that provides the design for the prototype falls into one of the following categories:

- Reused documentation (from the project design documents)
- Reusable documentation (to be fed into the main development cycle)

If the documentation that is produced serves only to support the prototype, then this could be seen as a questionable use of resources. The purpose behind creating a prototype is to address any issues that might arise as a result of implementing the design of the product.

Therefore, the prototype must, at the very least, provide proof that a concept works and how it might be implemented. In turn, this means that the work done as part of creating the prototype should be work that would otherwise have needed to be done in creating the product itself.

Thus, any documentation that is needed to create the prototype is also needed to create the final product. Two documents will need to be fed into the prototype phase, at a minimum:

- Requirements Definition
- Functional Specification

This last could be the Requirements Specification if the prototype is designed to feed itself into the Functional Specification and design for the final product.

To a certain extent, the documentation may also remain internal—that is, the client should be able to look at the prototype as a replacement for a formal design document that they should sign off as being a fair representation of the final product they want.

In such cases, it should be noted that, from the outset, the client is willing to accept a prototype in lieu of formal design documentation. It does not, however, replace the design documents that need to be created so that the programmers can build on the foundation provided by the prototype.

Indeed, they will need to have such documentation in place before they can begin to build the final product, and, if the prototype is acting as a kind of placeholder for that functionality, it may be required even before work on the initial prototype is begun.

THE PROTOTYPE AS A SKELETON

As we have seen, there are different ways in which we can view a prototype. Besides the third option of simulating a system prior to developing the solution for use in the real world, prototypes can be looked at in one of two ways—either as standalone products that serve a specific purpose, such as proof of concept, or as the start of the full product, a skeleton into which functionality can be dropped.

Games, for example, can be created without artwork in order to test the game mechanics and principles, to get the balance just right before the actual product is created. Office software can be constructed around a series of logically arranged menu options and dialog boxes.

Implementing Skeleton Prototypes

If the main reason behind creating the prototype is as a skeleton form of the final product, a few points need to be taken into consideration. For example, the non-functional requirements of the original design (such as operating environment) need to be respected at the earliest stages of the project.

These might have an effect on the decision whether to proceed with the proto-type as a skeleton, and the way in which the prototype is developed. They may influence areas such as programming language choice and platform, which will have an effect on the ease with which the prototype is to be created.

If a language such as Visual Basic is used to build the prototype, it will be some-thing that is quick and easy to create, but since VB is not known for its high per-formance, then if there is a nonfunctional requirement that dictates that the system performance approaches real-time criteria, VB would probably not be a good choice of programming language.

However, high-performance languages such as C++ and assembler bring an overhead in development time that will mean that the prototype will not be avail-able with the same immediacy as if a simpler language such as VB were to be used.

PROTOTYPE LAYERS

The previous two sections raised two important questions:

- How much functionality should be designed into the prototype?
- Should the prototype be the first step in a final product?

To answer these two questions it may be beneficial to think of the prototype in layers, or phases. Each can be as functionally complete as necessary to give an ac-curate view of the potential of the end product. If every layer of the prototype is complete, then the product can be said to be complete.

However, for reasons of stability, future expansion, and pure performance, the completed prototype may fall short of the nonfunctional requirements that the client has, and therefore a full product may need to be redeveloped, using the pro-totype as a living design.

The Interface

The first layer is that which is seen by the user—the interface. The interface can take a variety of different forms, depending on the product type. For example, should the product be an application, the interface is likely to be either a command line or a graphical user interface of some description.

However, should the end result be a library or component that is destined for inclusion in other projects, then it may have another kind of interface, such as an exported function that can be called to provide a service. Windows common dialog boxes (such as the File Open dialog box) are examples of this kind of interface, and are the reason why most applications have a similar look and feel.

GUI Interfaces

No matter what operating system is being used, GUIs usually follow a set pattern of features, which can be seen in Figures 13.1 (Windows), 13.2 (Menus), and 13.3 (Dialog boxes).

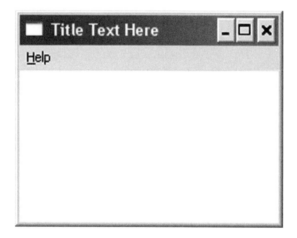

FIGURE 13.1   Windows.

It is quite possible to have a working application, which has all of the features in Figures 13.1 through 13.3, but without actually doing anything beyond supporting the interaction between the user and the application. A Visual Basic application of this kind can literally be created in a matter of hours (or minutes for a very small application).

This means that it might actually take longer to document the prototype than to create it, but so long as the documentation is reused, or can be reused, this is of little consequence.

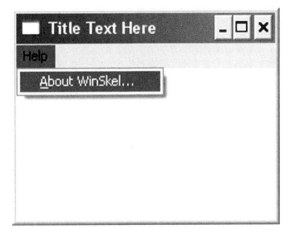

FIGURE 13.2 Menus.

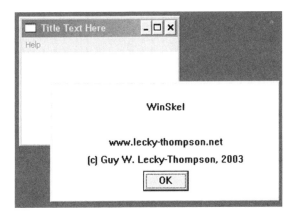

FIGURE 13.3 Dialog boxes.

Command-Line Interfaces

In the same way that the GUI indicates to the end user the possibilities of the application, the command-line options give clues as to what the application does, and its capabilities. Thus, the first step in creating a command-line driven prototype is to isolate the command-line options that will be required in the final product.

Figure 13.4 shows a typical splash screen, which shows how a specific command-line application is designed to be used.

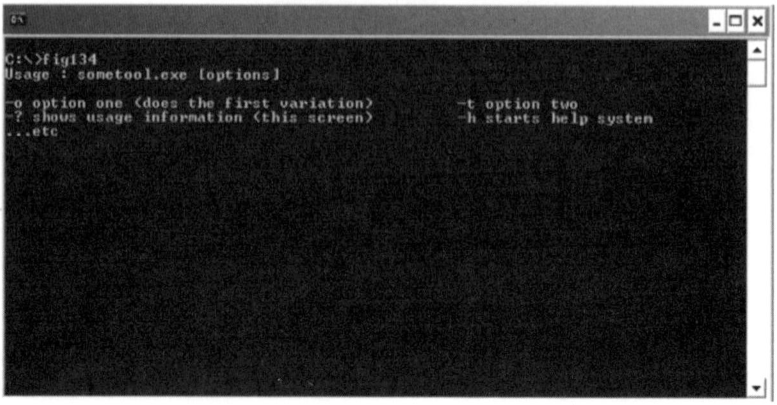

FIGURE 13.4   Application usage splash screen.

This is slightly different from the way in which such programs are usually constructed, which tends toward the "build and add" approach. However, in the context of software engineering, it can be argued that it is necessary to know exactly what the application can and cannot do before actual construction begins.

Hence, the command-line prototype will be able to demonstrate the capacity to process all the arguments on the command line before they are actually implemented, as opposed to creating a piece of functionality within the system and then attaching a command-line option to activate it, as can be the case in less well managed development processes.

Exported Functions

End products that either exist to be used as libraries to be imported into other products (language-specific object code or language-independent libraries) or as components (dynamic link libraries) to be used inside a larger system will only have an interface that is accessible by a programmer.

The exception to this is when the resulting component can be imported into other systems by an end user. Web-based projects are among those where the client might be requiring development of a component that is designed to be imported into another third-party tool by the end user.

In such cases, only the callable functions need to be exposed in the prototype—they will probably be filled with code that is designed to return a result that is of the kind that will be expected in the future, but without performing any real processing.

The Logic

The next layer down from the interface is the code that is designed to fulfill the requirements laid out by the client. In other words, it is where the bulk of the processing will be done.

Some projects will require that the logic is implemented in the prototype, by way of a proof of concept or proof of understanding. Other projects will require that there is just enough logical code behind the interface to illustrate that the interface is adequate for the purpose for which it has been designed.

Should the client require that more logic is implemented than this, it is important that the designers and programmers are aware that the code they write is potentially marked for reuse and thus must adhere to the standard practices of the organization laid down for code production.

External Dependencies

Finally, all applications need to be able to interface with other systems—whether peripherals, an operating system, or third-party device or network. These should be simulated for the purpose of the prototype except in cases where the product exists solely to provide an interface to such a device.

This also allows a limited amount of testing to be performed, especially in cases where the hardware to which the system has to interface is not yet available, or is very costly. However, there is a danger that the simulated hardware itself becomes a project in its own right.

The final point about simulating the interface to an external resource is that it enables the system to be built up much faster, and in a more controlled manner than if the connection to the resource needs to be made immediately.

THE DEMONSTRATION

Prototypes usually result in a demonstration stage of some description. That is, the prototype is delivered in person and demonstrated to the client. At the end of the demonstration, the client should be able to make a decision as to whether the developer has fully understood the complexities of their project, and hence be allowed to proceed and develop the rest of the system.

Active Prototypes

An active prototype is one that can be altered during the presentation process. In other words, the client can instruct the developer to alter the prototype such that it more closely resembles the look, feel, or logic of the model that they hold in their mind's eye.

This is a very useful technique, as it increases the extent to which the client is involved in the project and provides an easy way to change the way in which the developer has understood the client's requirements.

However useful it may be, realizing it requires that a development environment for the prototype be:

- Physically transportable (probably on a laptop)
- Easy to use/fast to implement changes
- Visual/Scripted

It is no use trying to provide an active prototype that is written in an obscure programming language, needs a mainframe to compile it, and takes half an hour of programming effort to change one aspect of the user interface.

The key to the technique is being able to involve the client, probably to the extent that they are allowed to change bits and pieces, understand the nature of the change, and see the results immediately.

For these reasons, it is usually reserved for the user interface aspects of a system, or the scripted behavior of prototypes that are a dry run for the final product and will need to be redeveloped in a higher performance language due to nonfunctional constraints.

ASSIMILATING CLIENT FEEDBACK

Prototype deliveries are usually accompanied by a contractual payment. The contract might state, for example, that 25 percent of the final agreed sum is to be paid upon acceptance of the prototype.

Passive Prototypes

If an active prototype is one that can be easily changed during the demonstration process, a passive one is the opposite. It is a prototype that has been built as a demonstration, in order to demonstrate understanding, but has not been created with the possibility for easy adjustment to measure.

This may be because it has been built with the same tools as those that will be used to create the final product, and as such, it is simply not efficient to correct errors in an interactive way. The most likely reason for this is that there is a non-functional requirement that prohibits the use of a rapid application development system (RAD) such as Visual Basic or C++ Builder.

Of course, it may also be because the investment would have been too great to create such a prototype, and so it has been thrown together using office tools for creating presentations, and graphic tools that provide the façade of an application, without any programming having taken place.

Whatever the reason, a passive prototype can only be changed outside the demonstration phase itself. As such, we would like to be able to reduce the number of iterations that need to be gone through before the final prototype is accepted, and the contractual payment made.

Pre-Delivery Conferencing

It is a good idea, if practicable, to try to let the client get their hands on a pre-delivery prototype, perhaps in a video conference or over a dial-up connection. This will allow them to identify major inaccuracies before the final delivery of the prototype.

It is important when using a passive prototyping strategy to involve the user in an off-line manner as quickly as possible. This will reduce the resource impact of producing an erroneous model.

This is a direct consequence of needing to develop a prototype that cannot be easily altered, due to the inherent complexity of the development environment or just because it is created with a simple, yet time-intensive tool.

Less good than an interactive pre-delivery session is the provision of screen shots and supporting documentation. It is unlikely, however, that a client will accept delivery of a prototype consisting only of documentation, since the idea is to also provide some kind of proof of progress.

Limited Redelivery

At the other end of the miniature life cycle that makes up the prototyping phase of the project is the delivery, and subsequent redelivery. Although we have tried, by using pre-delivery conferencing, to limit the number of errors that are present in the prototype that is delivered, it would be unrealistic to assume that only a single delivery need take place.

Therefore, a second delivery must be budgeted for, with the implication that if more than two deliveries are needed, there is a vital flaw in the understanding between the client and developer that needs to be addressed before any further work is carried out.

It is easy to get into a behavioral loop whereby the client is constantly reevaluating the prototype such that the developer feels that they will never be able to produce something that is acceptable. This is bad for the client-developer relationship, and is one of the key reasons why the principle of limited redelivery needs to be adhered to.

This may also be due to the client not fully understanding the role of the prototype in the development process. It is not supposed to be correct in every aspect, but needs to be sufficient evidence of understanding that the project can go forward, not hold up the subsequent phases because it is not exactly as the client intended for the final product.

RECORDING

Of course, the best note taker in the world will sometimes miss vital pieces of information, which is why it is often a good idea to record the demonstration meeting in its entirety, preferably using a video camera.

Simply recording the sound is usually not good enough, except as a memory jogger, since it is difficult to make the link between the spoken word and the image on the screen that represents the prototyped software.

The impact, however, of this is limited if the prototype is an entirely passive, noninteractive deliverable, such as a PowerPoint® presentation.

SUMMARY

This chapter concentrated on producing a piece of code that allows the client to be able to gauge the effectiveness of the developer, while at the same time requiring only a minimum of resource investment. The key aims of a prototype are:

- Proof of understanding
- Client involvement in development process
- Easy to validate system

To satisfy these, we need an environment that:

■ Is rapidly understood, flexible
■ Includes visual programming/scripting
■ Requires low system overhead

If we manage all of these issues, then a well-designed prototype, achieving the correct level of detail, will become a vital part of the development process.

14 | Adding Functionality

In This Chapter

- Introduction
- The Building Blocks Approach
- OO Development Revisited
- Unit Testing
- Of Menus, Glue, and Simulating External Dependencies
- Summary

INTRODUCTION

Since we are advocating an object-oriented approach, and one that begins with a prototype, adding functionality over a period of time by gluing in separate objects and components, we should also give guidelines as to how this functionality should be built in. It is not as straightforward as one might at first imagine, since each module needs to be created separately and then passed to a team responsible for integrating them into the final product.

This team will likely consist of programmers, testers, and scripters, all of whom expect a certain level of quality to be adhered to before the modules are submitted for inclusion within the final product. Not only does the functionality handed over to them need to be complete, and tested, but also well documented so that those who are using the modules do not need to dig down into the source code to find out how they are to be used.

In fact, the most efficient approach might be to actually retrieve the modules explicitly from the object and component archive that we described in Chapter 11, "The Object and Component Archive"; in this way, we can be sure that the code has been defined, implemented, and tested within the stringent quality guidelines laid out by following the practices described in this book.

To recap on the engineering process that we have thus far advocated, the following steps are usually followed to turn the design into a product:

1. Create prototype.
2. Iteratively add functionality.
3. Test entire product against agreed standards.
4. Deliver finished product.

In the previous chapter, we discussed the first of these phases, and in Chapter 5, "Testing," we covered testing, and the processes that govern it are also discussed in Chapter 17, "Testing Procedures." The next chapter deals with delivery of the product. Thus, it is the second step that we will be covering here.

THE BUILDING BLOCKS APPROACH

One methodology for creating the logic that will make up a large part of the end product is to connect a series of blocks together. These blocks represent pieces of functionality that fall into one of the following categories:

- Ready for use
- To be extended/adapted for use
- Must be created specially

Those blocks that are ready for use in the project, or that have to be extended in order to be used in the current project, represent reused objects that exist in the Object Repository. Those that need to be created specially for the project should eventually find their way into the repository for use in other projects.

There should be nothing substantial being developed that neither comes from the repository nor finds a place in it. Each piece of functionality needs to be created as a reusable object, or block, which could be woven into a future product.

Therefore, in following the building blocks methodology we are ensuring that we enrich the collection of blocks that we have available as projects are completed. This also means that the resulting code is more reliable since it has been tested in a wide variety of situations. It is complementary to the whole ethos of reuse, and object management.

Naturally, there are issues to address, such as the level of granularity we want to be applied in the repository, which will have an effect on the size of each block. If a product is created using many large blocks, each with functionality that is not used, we have probably chosen a granularity that is too large.

The real problem is deciding whether an object should be cut up into pieces, or whether two objects should be fused together to increase or reduce their functionality being offered. It is quite possible to maintain an Object Repository in which each object is reduced to its smallest possible size, but the result is that so much code is needed to glue the blocks together that the glue code itself becomes a source of quality defects.

One of the guiding principles of the building blocks approach is that we want to minimize the amount of new development that must be performed to create the product. The more new code we create, the more chances there are that the project will not be completed within the required timescale, or that errors will creep in and reduce the quality of the end product.

Sourcing Blocks

There should be only two sources for the blocks—the Object Repository or the Component Gallery. This implies a number of important points, the first being that if the code does not exist, and needs to be created, then it cannot be used until it has been incorporated in the Object Repository, or purchased and placed in the Component Gallery.

The next important point is that programmers working on the project exist to take blocks and connect them together to produce the final product. They should not be writing code that has any substantial logic to it. This is not to say that they would not find themselves building such code on behalf of another project, but rather that they should not be doing so for their own project, except under certain circumstances.

In essence, this also means that the job of project programmers becomes one of research and incorporation, rather than creation. If the block they require does not exist in the Object Repository, and they cannot find an appropriate piece of code in the Component Gallery, they have two choices.

They can either schedule the development of such a block (object) via the Liaison Center, or look for an Open Source component that can be incorporated into the Object Repository. They might find themselves developing the object, or they might equally find that there is some spare capacity elsewhere that can be taken up by developing the functionality for them. It is also possible that the creation of the object will be outsourced to another organization entirely.

The defining boundaries of each block will be specified by the design of the system under creation, and the philosophy by which the Object Repository has been created. Blocks may well come in different shapes and sizes, each offering functionality that can be reused, and sometimes, like a farmer building a wall, it will be a case of choosing the right block for the purpose.

While the building block methodology might seem like a long-winded approach to writing software, one has to measure it against the gains in quality. It is worth the investment in time that will be required to find and integrate an object, rather than trying to develop it from scratch.

Part of the reason is that programmers often find that, once they have begun development of a specific object, there are issues that need to be addressed that they had not thought of, which they discover as a result of actually doing the work.

Therefore, in developing an object from scratch, programmers may well find that they had underestimated the problem, and that it will take longer than they had first anticipated. More complexity means more code, which also means more room for error, and before long, it becomes clear that, had the reusable object been found, it would have been more sensible to use it rather than to code the object from scratch.

Cross Coding

Part of the success of the methodology is reliant on the principle that each block is well defined, implemented, and tested. If the object that is required lies outside the scope of common knowledge, it must be developed internally or outsourced.

The two solutions are, to all intents and purposes, identical in approach, if not in cost—it will probably cost more to outsource the object development, both in time and money, than developing it internally, but the way in which it must be created will not differ.

Cross coding is a term that describes the fact that the team developing the solution is not the same as the team creating the various parts that make up that solution. Part of the principle is that in developing in this way, a third party is required to understand the problem, and create and test a solution, based on knowledge passed on to them by the project team.

This is identical to the approach that the client is taking with the developer—they also have to communicate their wishes to a third party to obtain a solution that caters to their specific needs and that can be verified with reference to an agreed set of criteria.

Therefore, it is one of those cases where an internal client is involved; the relationship management is identical, and is still performed through the Liaison Center and other channels that have been set up for those purposes.

Of course, we expect that there will be aspects that are carried out more efficiently, since it is a dialog between two technical parties. However, since a certain amount of nontechnical requirements will also be discussed, the hurdles and goals may well be very similar to those that govern the overall client-developer relationship.

The term *cross coding* refers to the fact that programming teams will find themselves working for each other on projects with which they may not be familiar, as if they were outsourced developers. Ideally, the team requiring the development work should not be aware of who is providing the code—an internal or external source.

In this way, it is hoped that adherence to strict guidelines and quality controls that need to be in place to govern the relationship between client and developer will yield higher quality code than if the project team were to develop their own code, and simply insert it all into the Object Repository at the end of the development cycle.

By using a different paradigm, the code will be placed in the repository before use, thus effectively ensuring that all code used by the project team is of the same quality. The added impetus for project teams involved in creating the code is that they may very well end up reusing the object themselves in the future and should therefore be extra vigilant in making certain that it is of sufficient quality that they themselves would choose it for use in their own projects.

Iterative Development

We mentioned that each object that is to be used in the project should be developed as if it were a little project in itself. In essence, this means that the specification, design, implementation, testing, and delivery phases all need to be respected, as do the corporate guidelines with respect to documentation, coding style, and quality assurance.

It is acceptable for each of these objects to be treated as little Waterfall model software engineering projects, with all the restrictions that this, more traditional, approach might entail. We can reduce it to the minimal approach in Figure 14.1.

Of the five steps in Figure 14.1, two of them require interaction with the project team, so it is, in practice, unlikely that we will be able to maintain the kind of separation that is possible when the two parties do not share an organizational relationship.

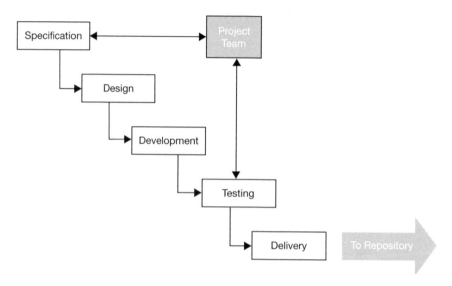

FIGURE 14.1   Object development process.

Nonetheless, this should not change the fact that the project team requesting the object development is a client, and should be treated in the same way as any external client. After all, the net effect will be that the code is delivered to an external client, and so the same level of care needs to be taken over its development.

Strictly speaking, this communication should take place through the Liaison Center, but, depending on the size of the target organization, this may not be entirely practical. Larger organizations, with many teams and a variety of projects under development at a given time, may find that it is a workable approach, which should yield higher quality results.

The Design and Implement stages can be performed in isolation; that is, there is no need to report back to the client every step of the way (in this case, we are referring to the "internal" client). However, the object needs to be tested in accordance with the requirements of the client, so they will need to be involved at the Testing stage.

The last stage requires that the tested object is released into the repository, where it can be used by any other project team. This stage, at the very least, should be performed with the Liaison Center so that the relevant librarian and repository management tasks can be performed to maintain the high levels of quality throughout the organization.

OO DEVELOPMENT REVISITED

Even though we have done our best to separate the roles of object developers and project team developers, it is inevitable that a slightly different approach needs to be taken when creating objects for use in a larger product. This applies to internal clients as it does to locating third-party objects that might need to be adapted before they can be placed in the Object Repository.

Specification

The objects that are being implemented during the process of fleshing out the prototype so that it offers the desired functionality are simple enough that they can be referred to in terms of a mapping of inputs onto outputs. In some cases, it may be enough to specify the inputs and outputs without describing the mapping in specific detail.

The documentation that is created when the specifications are written can then be reused later when an object description is needed so that future projects might take advantage of the specific features being offered by it.

It may also be the case that all that the system developers have in hand when they implement the object (or reuse the object) is this specification, so it needs to define exactly what the object is supposed to do.

We should reiterate that the point at which we begin to perform this kind of task is the point at which we have broken down the entire system into easy to implement units. Thus, there should be no need for any detailed design guide to the object because it must be at a reasonably low level of granularity.

However, that is not to say that an object will have only one mapping of inputs to outputs, and we need to bear in mind that additional state information might be needed in the specification of the object. At this point, it begins to resemble more than just a simple case of mapping inputs to outputs, but at the lowest level of granularity, this is what we are trying to achieve.

The end result might then be that we have a series of functions (or methods in OO parlance) that are all defined as a set of inputs mapping to outputs. The output can be information returned to the requesting object, a change of state of the object under specification, or a mixture of the two.

There might also be some private object functions that will need to be specified to make sure that the implementation team knows exactly what the object needs to do. The team responsible for this implementation might not even be members of the original project team, so it is wise to be as explicit as possible.

If the rapid application paradigm requires it, there is the possibility that known mappings can be used, and hard coded. These kinds of placeholder functions will always map a given set of good known values onto a subset of values, and the result is that the object can only exist in a limited number of states.

Assuming that all the states of all the objects contained within the system can be proven compatible, for the limited number of cases that the slowly evolving system is supposed to be able to deal with, then this may prove enough of a demonstration for the development to be completed.

If a more iterative approach is being used, each function will gradually become enriched, with new features added to it as required until the system is in a state in which it can be delivered. The paradigm that we have chosen in this book is flexible enough to accommodate almost any approach to fleshing out the prototype, as we concentrate more on the process than the actual practice of building software.

Testing

Part of the reason why we break the system down in this way, under the OO paradigm, is that it makes it much easier when it comes to testing. Since we have a discrete set of inputs that can be mapped onto a discrete set of outputs, it becomes easy to predict what an individual object will do in a given situation.

We need only know the inputs, outputs, and internal state to know how it will perform when something, either expected or not, happens in the system. Since we know what each individual object is capable of, and the way it works, it also becomes easier to be able to know how the entire system will react when it is put together.

This is one part of the way in which we try to reduce the problems that are associated with what we have previously termed the *intangible nature of software*. It becomes very difficult after a given number of collisions of balls on a table to predict where they will all end up, and software as a collection of objects has something of the same nature, which is why we concentrate on testing everything at least twice.

Since we have specified what the objects are supposed to do in terms of a set of inputs mapping to a set of outputs, the actual testing process, as we shall see, becomes rather easy and well defined, at least at the object level. However, as we start to put the system together, it becomes very difficult to be able to say, without building a simulation of the system, whether it will react in the right way.

This is the point at which we realize that we need to be able to rely on the fact that the individual objects have been well tested; otherwise, we will not be able to effectively locate the source of the problem. Therefore, good specifications and good testing will help to drastically reduce the possibility of finding errors in the final system, and reduce the time taken to locate the source of the error and, eventually, fix it.

UNIT TESTING

While we have covered the subject of testing, or at least the mechanics of it, before, now we should look at it in terms of the current discussion—adding functionality to an existing framework by the integration of objects. There are two kinds of objects that will need to be integrated—those we have written, or had written as part of the current project, and those that we have acquired as a result of a search of the Object Repository or Component Gallery.

We need to test in phases, starting with the smallest unit and gradually adding objects and the code that glues them together until we have adequately tested an entire functional area. At this point, that piece of functionality, perhaps linked to a menu or other user interface artifact, can pass into system integration and integration testing.

It is vital that the processes that we are about to define are linked in with the solution that the organization has chosen for managing the problem report and change request cycle so that, in the event of an error being located, the problem can travel down the return path and the solution be fed back into the development process.

Functions

The smallest unit to be tested is the function. This may be a standalone function or part of a library of functions that provide a specific set of services, or it may be a function developed as an extension to, or simply part of, an object that we would like to use in the system.

As we have discussed, functions need to be well specified, in terms of a discrete set of inputs mapping to a discrete set of outputs. Only then can we begin to test the function in an efficient way. The specification can either be a separate document or it can be derived from the source code. The latter has advantages in that it enables the documentation to be kept with the code in an easy manner, but depending on the implementation of the Object Repository, a separate document might be more appropriate.

The first step is to create a test data set, as outlined in previous chapters, based on the mappings from input to output. Since we have defined a discrete set of mappings, it should be quite easy to create at least a general set, if not specific use-case style examples.

With our test data in hand, we can then proceed to test the mappings by feeding in inputs and looking at the outputs. At the very least we should perform Weak Testing, if the function is an extension, but preferably adopt a Strong Testing approach, the difference being that Strong Testing is designed to verify exception cases in addition to use-case style tests. These are examined in detail in Chapter 17, "Testing Procedures."

The findings need to be reported, either back to the team responsible for the implementation should the results not be satisfactory or forward to the project manager so that the next phase can be considered, should they be acceptable. The project manager could also intervene to accept slightly less than perfect results if he decides that the impact of the failed tests will not affect the use of the object in the resulting system.

There is no need to test reused code, however, since we assume that it has been tested as part of the introduction cycle into the Object Repository or Component Gallery. The use of the object may be slightly different, and the results of the previous testing process need to be examined, but we need to be able to have faith that the code that is returned from the Object Repository or Component Gallery is fault free.

This approach will save time in the development process, as long as the paradigm has been followed properly and no corners have been cut. This means that the entire organization needs to be appraised of the importance of good quality development.

Objects

The next level up from a function is an object, which can be seen as a collection of interactive methods, both private and public, coupled with the internal state of the object to which we might not be privy. Unit testing objects is a precursor to binding them together into functional areas for further testing.

Again we need to verify that the behavior is correct, based on the specifications that are available for that object. It is on the basis of the specifications that we have created, had created, or chose to reuse the object that is being examined. Therefore, any testing that is performed, be it to validate an extension to the object or to test previously untested behavior, will be on the basis of the specifications.

To do this we need to construct a set of meaningful test data, which could be based on previous, historical testing sessions, or it could be real use-case style test data that we have put together based on the anticipated way in which the object will be used within the system.

Here again, we should mention that it may not be necessary to actually test objects that come from the Object Repository or Component Gallery, if we have not altered them in any way since they were previously examined.

Again, this is a way of making the whole development cycle more efficient, but it does mean that we need to be able to rely on the quality of the objects, based on the implementation of a correct paradigm for the creation and control of the processes surrounding their creation.

If a problem is uncovered in the additional functionality that we have added for this project, the code needs to be returned to the team responsible. If a problem is located that is the combination of a function that was already existing, plus one that has been added, then the specifications and testing data need to be examined in an attempt to find the true cause of the problem—it could be the specification, design, or implementation. It is for this reason that we insist on the functional units being as small (in terms of implemented logic) as possible; it makes predicting their behavior in a group much easier.

Functional Areas

The final part of Unit Testing is to test interacting objects in a specific functional area, which could be linked to a user action or a system request. It is not likely that, in the system, something occurs that is not the result of an action or external stimulus, and so breaking down the system into functional areas, which we first did in the Definition and Specification processes, should be reasonably easy.

At this point, all objects in the system will need to be tested to ensure that they operate together in a correct manner. The code that connects the objects together should be simple enough as to not require extensive testing; the main reason for performing functional area testing is to be sure that all the objects can work together in a way that is conducive to providing the required functionality.

If an error is located in the code, the code needs to follow the return path back through object testing to functional testing in an attempt to understand what mapping from input to output failed. There is also the possibility that the internal states of the objects are not correct with respect to their specifications.

The test data that is used to test this part of the Unit Testing is based on an analysis of the problem domain. That is to say that the test data needs to reflect as many anticipated interactions as possible.

Once this phase is complete, the functional area has been tested, with respect to the objects that are present in it, and it can be passed over to System Integration, where it will become part of the larger system, by connecting it to the other functional areas, perhaps by using further glue code, logic, or based on user interface artifacts, such as menus or dialog boxes.

Until the system integration phase, the functional areas, and the objects that exist within them, are isolated from each other and the rest of the system, and it is not until we pass into Integration Testing that we will know whether the system actually works. However, there is no reason to assume, considering the care that has been taken to match the specifications with the behavior, that this should not be the case.

OF MENUS, GLUE, AND SIMULATING EXTERNAL DEPENDENCIES

The final step in adding the functionality that makes up the system is to collect all the functional areas together and prepare the entire system. It is the point at which we have the interface, the objects in the functional areas, and we need to connect all the different pieces of the system together.

It is essentially equivalent to System Integration, and ends with Integration Testing, after which point we can say that the system is ready for final delivery, and Acceptance Testing if required.

The whole philosophy, when followed, results in higher quality software, and fault-free software. There is a subtle difference between these two in that high-quality software may meet the requirements of the end user without being entirely fault free, although we would hope that fault-free software would also be considered high quality.

The User Interface

The prototype sets the pace for the look and feel, and during the integration process, it is the starting point for building in the logic that fills the gap between the actions of the user and behavior of the system. The functional areas that are linked to the interface might work in isolation, but it is only when the system becomes connected together properly that any potential conflicts between functional areas will be noted.

The various dialog boxes and menus will already be in place, as well as any other user interface elements that will be specific to the project. If the system does not have a graphical front end, if it is a collection of command-line tools, for example, then the interface needs to be tested on this basis rather than trying to invent any other kind of testing process, or leaving it out entirely.

The client must accept the UI before final delivery, and preferably before the prototype moves into the system integration phase. This is so the functional areas can be examined properly in a way that makes them at once easier to analyze and test, and so the client can validate that they have been well understood—they will find it easier to do this if they have something to relate to.

Glue Code

Anything that offers no immediate logic can be considered glue code. It binds the user interface to the dialog boxes and the objects that actually implement the logical areas of the system to provide the functionality that the client originally requested.

Essentially, it is used to communicate results of objects' actions to other objects, and actions from the user interface or system to the objects by functional area. It is used in the preparation of the units to be tested and in actually connecting the other parts of the system together.

The general rule to apply is that one should try to use as simple coding as possible to reduce the likelihood of errors creeping in that are not part of the functionality of the system. The reason behind this is that the glue code is likely to be the least tested of all the different areas of development and design.

The aim is that glue code must be subjected to as little testing as possible due to the inevitable time constraints that start to play at the end of a project. We should not forget that the process of putting all the elements together is the last time that we can correct errors before the client appraises the product.

However, we should have tested the glue code by functional area to be sure that the conduit is working correctly, and tested it with relation to the user interface. What we have not spoken about yet is how it is tested with relation to the system or other pieces of hardware or software with which it needs to interact.

External Dependencies

Any specialist hardware or software not available at the time of creation needs to be simulated, and this will include the operating system, peripherals, or software applications such as backend processing databases. On the one hand, we cannot wait until they are available to begin the testing or integration process, but on the other, there will be possible problems that stem from using inadequate simulations.

These will most likely be problems relating to not having tested the simulation properly, it being another piece of software that makes up the system. Even if it is not delivered as part of the system, it is still something that creates dependencies within the system.

Of course, as the organizations become more expert in their field they will build up collections of simulated entities in the same way as they build up an Object Repository or Component Gallery, and they may also be able to acquire simulated entities through other means.

There is an inherent trade-off between simulating an external system, which requires that software is created to do so, and waiting for the real devices or applications to become available. Introducing another dependency that has the same complications as the actual system itself is not without risk, since it requires that an entire development process is started for the purpose.

However, it can be hard to test a system without access to either simulated units, or real units that provide the correct behavior, so in some cases it might be necessary to perform the additional work to ensure that the system will correctly integrate with the user's environment.

The problem can be partially solved by using strict Object-Oriented style input to output mapping and Specification and Unit testing techniques that we have described previously. This will rely on there being adequately expressed specifications that are also correct and define the way in which the external components have been implemented and the behavior that is expected from them.

SUMMARY

This chapter took us on a brief journey through a part of the software engineering process that is worthy of a book of its own. It is impossible to describe the many different programming approaches that exist within the programming community, and the various safeguards that individual organizations have put in place to ensure that the development process is adequately applied.

In fact, all we did here is suggest a framework that is compatible with the way in which we expressed the processes should work to try to promote a continuously improving collection of objects to minimize the appearance of faults within systems that are created.

The reader is encouraged to find a way in which the programming itself, and the process of creating glue code to hold all the elements together, can be performed in a fashion that is in tune with the corporate entity, and understood by the programmers who will actually be developing the code.

There are more and more paradigms to choose from, some that are cutting edge, and only now receiving acclaim, such as Extreme Programming, and time-tested ones that use simpler processes for the conversion of complex requirements into well-constructed code. The choice of languages used and the tools available to manage the development process will also play a part.

The discussion we had here should be compatible with all paradigms and tools for developing software, and was written in a way to describe the processes and not the implementation of them at the ground level, which is a guiding principle of the discussions led by the book.

15 Delivery

In This Chapter

- Introduction
- Preparing the Application
- Supporting Documentation
- Additional Customization
- Training
- Summary

INTRODUCTION

Finally, the product will be complete, or at least development will be stopped and the product handed over to the client. It is always desirable to deliver a complete product, but it is also a fact of development life that most projects are delivered lacking certain functionality that was in the initial design as they run over budget and are delivered late.

Of course, if every piece of advice in this book is followed, there is no reason why the situation should not improve over time, as the object and component archive becomes populated with high-quality reusable components. In fact, the Utopia that is object-oriented development, combined with efficient code reuse, should eventually result in products that are delivered on time, within budget, and with all the functionality that the client originally intended.

Therefore, the final phase in the project will be the delivery of the product to the client, and hopefully payment of any outstanding monies to the developer by way of confirmation that the contract has been upheld by the developer.

The package needs to consist of tangible and intangible deliverables:

- Documentation
- The application
- Proof of conformance
- Installation and customization training

Each of these is discussed in detail in this chapter, with the aim being that every project follows the same cycle, which serves to concrete the procedures so that with each iteration, they become more efficient and yield ever-higher quality results.

PREPARING THE APPLICATION

The relief that the project has come to a successful conclusion (after all, the application has been tested) often leads to the misguided conclusion that all that remains is to copy it onto a CD (or diskette), send it off by recorded delivery to the client, and await payment.

The reason is that the actual delivery is seen not as the culmination of the effort of the entire project, but a step that is necessary afterward in order to get paid. This attitude changes from the consumer industry to the on-spec industry that we have looked at here.

Those working in the consumer industry understand that the delivery and preparation of the delivery are as important as the design and development of the application. If it is not performed well, the chances of selling enough copies to have made the development cycle worthwhile are slim.

There is no reason why on-spec projects should not follow the same process in the preparation of the release as commercial projects; after all, the satisfaction of the client is at stake.

Media

The delivery media should match the target platform, and size of deliverables. This is obvious, and like so many other aspects can be overlooked. It should never be assumed that the client has more than the usual array of hardware at their fingertips. If in doubt, it is always worth asking.

There is nothing worse than preparing a DVD-only release, simply to find that the client does not have a DVD drive in their computer to read the disc. By a similar token, if the product must span multiple pieces of media, then a double-check to ensure that they have all been created correctly should be part of the pre-delivery checklist.

Installation Routine

The deliverable should be self-installing. While it is true that almost everybody has a piece of decompression software on their systems, it should never be assumed that the one that the developer uses is the same as that which the client has available.

Consequently, the developer should provide either an uncompressed deliverable or, at the very least, an application as part of the delivery that is able to unpack the contents of the delivery media automatically.

There are software products in the marketplace that can be used to create user-friendly installation and uninstallation scripts that should be used to automate the installation process as much as possible. Even if the product is designed to be installed by a field engineer or developer representative, it should be packaged such that the installation process is painless.

The reasons for this are first that there is every possibility that the client will have to reinstall the product at some point in time, and they probably do not want to have to call the developer when it happens.

Second, there is nothing worse for the image of the company than an installation that fails, even though it is being performed by a professional. Using a commercial tool to create the installation package minimizes the chance of a failed installation, since the tool can automatically detect dependencies and install appropriate add-ons.

Modern operating systems, including Windows, Linux, and many of the established Unix-based operating systems, have their own installation routine management software packages that form part of the operating system deliverables and can be used to package the installation for the product being delivered.

Assuming that a suitable tool can be found to interface with the operating-system-provided installation management application, it is always best to use it, as it will guarantee that the end user has it installed and is able to make full use of the delivered package.

Pre-Delivery Testing

To perform this testing properly requires a clean machine that has been installed as close to an out-of-the-box specification as possible. This means that the operating system must be installed on a system that has had its hard drive reformatted.

Then, the package needs to be installed, and with any luck, as long as the operating system is the same, or earlier than that on which the package is designed to operate, the test installation should reveal any flaws in the installation process.

This is important because the systems used to develop the product will have a very different profile from those that exist in the field. They will have all manner of tools and libraries installed that may be required by the product, but may not be part of the underlying operating system.

If there are such instances of third-party libraries or tools that are not present on the target system, but are required for the operation of the product, the developer is responsible for acquiring the permission of the original developer to redistribute these parts.

For example, libraries that offer certain additional functionality and are delivered as part of a compiler toolkit or development environment are probably freely redistributable. They should be listed as such in the accompanying documentation, under the title "Redistributables."

The developer may choose to make this a policy decision, such as requiring all software to be statically linked, whereby these libraries are incorporated with the final product. This means that, for each piece of software that needs the libraries, the code will be duplicated, since the libraries will form a part of every delivered product.

However, it gets around the issue of needing to get redistributable permission, and avoids version conflicts with existing systems.

One final point to add to the pre-delivery checklist is to test all the links to help files—both those that are context sensitive and those that are accessed from the menus. It is one of those aspects of the system delivery that tends to be overlooked, until the user requires the help file and finds that it has not been shipped with the application software.

Delivery Overview

Each piece of deliverable needs to be listed in an overview document to ensure that the client is able to verify that they have received a complete package. This includes:

- System Requirements
- Media Contents
  - Application Code
  - Third-Party Components
  - Permissions for use of third-party redistributables, where appropriate

- Documentation
  - Installation Guide
  - User Guide
  - Miscellaneous Documentation
- Hardware
- Passwords/Usernames
- Invoice

There should also be a cover letter, which is a polite way of introducing the package and lists the physical contents. The entire package should be delivered in some form of folder or binder that is designed to keep the media and associated documentation together, as well as any hardware other than large peripherals.

This hardware, as listed in the Delivery Overview, can be anything ranging from a copy protection device (or dongle), to large pieces of specialized equipment necessary to provide the functionality required by the client.

Security Mechanisms

Traditionally, copy protection has been used as a way of ensuring that heavily customized software applications or commercial systems cannot be illegally copied and used on a platform other than the one for which it was originally licensed.

In other words, if the client purchases five licenses, then they are allowed to install the software on five machines, and there needs to be a mechanism in place to ensure that this is the case.

The industry standard way of doing this for expensive, noncommercial but otherwise at-risk software has been through the deployment of copy protection devices known as *hardware keys*, or *dongles*. Without the dongle, the software will not run.

Software that is not at risk is that which has been developed for hardware that the average client has only one, or a limited number, of. For example, the manufacturer of missile guidance systems is probably not too worried about the implications of fraudulent use, whereas a company marketing a customizable, scripted accounting system that runs on standard hardware probably will be.

For use with PCs, there are two main devices:

- USB
- Parallel (printer port)

Essentially, the procedure is simple: if the dongle is not there, the software will not run. There are several key things than need to be checked, which may have an effect on the type of copy protection used:

- The target system may not support the connection required.
- The target system may employ another, noncompatible scheme.
- The target system may not be supported by the dongle manufacturer.

Finally, the developer needs to make sure that the required libraries, if any are needed to use the dongle, are delivered to the client, along with any activation codes, and a separate installation procedure to ensure that the software, once installed, can be used.

SUPPORTING DOCUMENTATION

In the previous section, we mentioned that some documentation plays a vital part of the delivery:

- Installation Guide
- User Guide
- Miscellaneous Documentation

These three areas of documentation are vital, since without them, the client will be potentially unable to use the product. Additionally, having these documents prepared and shipped will help with knowledge retention on both the client and developer sides.

Installation Guide

The aim of the Installation Guide is to take an inexperienced user through all the various steps that need to be taken to install the software, while also enabling more technically competent users to perform the installation without making any incorrect assumptions.

This might need some explanation. The issue is that technically competent users, when faced with an installation guide aimed at a reasonably inexperienced user, will tend to skip parts that seem familiar, assuming that the software installs in the same fashion as other pieces they may have installed in the past. In doing so, they might miss important information.

Therefore, the developer needs also to provide a Quick Start Guide specifically aimed at these types of users. The full list, therefore, might look something like this:

- Pre-Installation Checklist
- Quick Start Guide
- Step-by-Step Guide
- Security Device Installation
- Troubleshooting

The Pre-Installation Checklist ensures that the users verify that all the media, hardware, and associated pieces of documentation are available before they begin the installation process, as well as checking that their target system meets the minimum requirements for installation and use of the software package.

We have already dealt with the difference between the Quick Start Guide (experienced user) and Step-by-Step Guide (novice user), but there should also be a reference in both of these to the next item—Security Device (copy protection dongle) Installation. Usually, this has to be carried out prior to installing the main software package. The installation guide provided by the manufacturer of the dongle should be used.

Finally, the Troubleshooting Guide needs to provide answers to any questions that might be raised during the installation process. As a minimum, all error messages need to be documented here, along with details of how they can be resolved, if resolution is possible.

User Guide

The User Guide is an important piece of documentation, and should be written by a technical writer, analyzed by a member of the nonprogramming team, and handed over to the client with the expectation that there will be pieces missing.

A User Guide is never complete, by virtue of the fact that it is almost impossible to plan for every contingency. There will be an almost infinite number of possible installation circumstances, and users can sometimes try to perform operations that were not foreseen by the developers of the system.

There will also be things that happen during the normal operation of the software that cannot be foreseen, such as system crashes, resource crises, and so forth. The User Guide can only deal with the normal operating conditions, and those exception cases that are known ahead of time and planned for. Anything else will result in a call to the help desk.

Theory of Operation

This part of the User Guide should aim at giving an overview of the entire system and the problems that it is designed to solve or features that it offers to the user. This overview should then be broken down by subject area in order to detail all possible options available to the user.

It is reasonably easy to prepare a skeleton document for those applications that have a graphical user interface by simply ensuring that there is a section for every menu item and dialog box that the application developers have created.

This information should also appear in the design documentation, which will enable the technical writers to begin developing the outline before the application is fully developed.

Command-line utilities can be examined in the same manner, but with each possible command-line option being examined, as well as their interdependencies. There should also, for command-line utilities with more than three options, be a series of tables that detail the various allowed combinations of command-line options and flags.

Any additional text or binary files that are used as input to the system (be it a command line or GUI application) also need to be detailed in this part of the User Guide, with the detailed file format as part of an appendix.

Task-by-Task

This part of the guide deals with the same subject matter as the theory of operation, but from a task-oriented point of view rather than an examination of every possible menu option, dialog box, or command-line option. Of course, there will also be references to those pieces of the interface that have been detailed in the Theory of Operation.

It is important that the Task-by-Task Guide is aimed at a novice user, since it is probably going to form part of any on-the-job training that the client embarks on as an alternative to formal training plans offered by the developer.

Troubleshooting

Like the Installation Troubleshooting Guide, the starting point for this should be a list of all the possible application error messages, and an explanation of:

- Why they have occurred
- What the user should do immediately
- Whether the user can perform remedial action

It is reasonably easy to know what application defined error messages exist, but probably more difficult to anticipate the equivalent for operating system generated messages that result from use of the application. Nonetheless, any that cropped up during system testing should be noted in the Troubleshooting Guide.

There needs also to be a section dedicated to resolving issues that are found by virtue of the fact that an expected external operation did not occur. This applies to things like interaction with peripherals not being handled correctly, loss of data, and so forth.

Of course, once the product is used in the field (assuming that it is not a one-off), users will feed back problems that can be included in the Troubleshooting Guide released with future versions.

Problem Report Procedure

Rather than leaving it up to the user's imagination as to how a problem with the software should be reported, it is best to detail exactly how it should be done as part of the supplied documentation. The procedure and supporting documentation used should reflect the standards used by the developer across all projects.

What follows here are simply suggestions as to what kind of procedure could be adopted and what documentation should be created to support it. It is important to note that the person reporting the problem might be the end user, and not the client; in such cases, the user may not be able to distinguish a real problem from an issue that arises due to improper use of the software application.

Therefore, it would be useful to include a preparatory note that informs users that they should only initiate a Problem Report if they are experiencing improper behavior of the system that cannot be explained by consulting the User Guide.

Forms

In order to provide support for structured reporting, it is necessary to provide the user with a set of forms that can be used when informing the developer of a suspected problem. The information that should be included could be:

- Scale of Problem
- Operational Area
- Menu Option/Dialog Box
- Sequence of events
- Hardware Platform
- Software Version

In addition, there should be a paragraph explaining the various ways in which the form can be submitted: e-mail, fax, regular mail, and so forth. This may be linked to other items such as the Scale of Problem.

Scale of Problem

This can be categorized in a variety of ways; for example:

- Blocking
- Nonblocking
- Cosmetic

These three key problem definitions essentially mean that the problem is preventing use, or it will prevent use at some point in the future, or is something that should be fixed because, while the application still works, the user interface does not match the actual behavior of the system.

Operational Area/Menu Option/Sequence of Events

This can be arrived at by essentially asking the user to repeat the sequence of events that led to the problem, in the hope that:

- It is not repeatable, and probably a result of user error, or
- It is a real problem, and can be repeated by the developer

Asking the users to provide the exact sequence of events, along with the menu options clicked and the data entered into any dialog boxes or at any other prompts (command-line systems) ensures that they think about the steps they have taken. In doing so, they might just find that they are doing something wrong.

The worst kind of fault is one that cannot be repeated by the developer, or even consistently by the user. If the system works at least some of the time, it will give the user some breathing space to try to resolve the error in partnership with the developer before it becomes a blocking problem.

Platform Information

The software and hardware platform needs to be specified, along with the version of the system being used. It might be beneficial for the developer to include a self-documenting menu option that automatically generates a report that can be forwarded to the developer.

Common information might be:

- Operating System
- Hardware manufacturer
- Processor speed/Amount of memory/Free disk space
- Software major/minor version

There may be other pieces of information, such as those relating to any device drivers or additional hardware that the developer might require. These should also be explicitly listed on the form.

Response Time

The response time will vary depending on the severity of the problem, and complexity of the solution. What is important is that the client is aware of and accepts the response times that the developer indicates, and that the developer actually respects the response times that they communicated to the client.

There should be an initial response confirming that the problem can be repeated, and that the technical team will look into it. This needs to be followed up by a report as to what the developer intends to do about the problem.

An interim solution (or workaround) should be offered at this stage for blocking problems, since without it the client can no longer continue working. This should be communicated within a reasonable amount of time. Exactly what is meant by *reasonable* will depend on the industry—nuclear power stations and other real-time systems probably work to a slightly different set of rules regarding urgency than small businesses using a simple accounting package.

The final solution should then be identified, and an appropriate plan for developing, testing, and delivering the solution worked out internally and then presented to the client.

There may be cases where the client claims a problem has been found, but the developer finds that it is due to a misinterpretation, ambiguity, or incorrect statement in the specifications or design of the product. In such cases, the client may be asked to intervene in the cost of the solution; however, third-party arbitration with respect to the legally binding contract will probably be needed to arrive at a mutually beneficial agreement.

ADDITIONAL CUSTOMIZATION

In the life span of the development of a product, there may be cause for the client to request that additional features be added to the system. It may also be the case that there are aspects of the system that can only be determined once it is installed at the client's site.

User Customization

This is an optional part of the User Guide that is only required if the product is designed to be customized by the end user, either immediately or at some point in the future that is not defined when the product is delivered.

It should clearly indicate which parts of the system can be user customized, and which parts cannot, as well as give an indication of the level of support that the user can expect once the product has been customized.

It may well be that the developer decides to give no free support whatsoever for those parts that have been or can be customized, including the possibility that if the user changes a piece of the system such that it causes another, noncustomizable part to behave in a fashion that is detrimental to the entire system, then the developer is no longer responsible for the damage.

Change Request Procedure

In the same way that a procedure needs to be defined to deal with cases where a problem is located in the system, there should also be a specific procedure that deals with developing new features or changing existing ones.

In this case, the procedure becomes similar to that which was used for the initial product development:

1. Client requests change.
2. Specification developed.
3. Change implemented.
4. New product is tested.
5. Product is delivered.
6. Client validates change.
7. Client pays developer.

The difference between this and the problem-reporting procedure is that the client is much more involved, since they will be paying for the change. A stage will need to be inserted in the preceding list, in which the developer and client decide on a price for the implementation of the change with respect to the complexity and the added complexity it brings to the entire system.

There is a difference from the original development cycle too, in that the client is potentially much more involved with the creation of the change specifications and the testing of the new functionality than they were with the original development.

One reason for this is that they have experience using the system, and this will leak over into their willingness to try to help the developer create a better product, or at least one that more closely matches their needs.

This will not be the case for products that are aimed at the commercial marketplace, where suggestions made by users for new features are usually welcomed, but unless a special relationship between the user and the developer exists, this is the beginning and end of their involvement in the process.

Effect on Maintenance

For those projects that are produced on spec—that is, specifically for a single client, or are extensions of off-the-shelf products—any changes will have an impact on the annual maintenance cost of the system.

This is because the complexity of the system will change as a result of introducing new functionality or changing existing functionality. The extent of this impact needs to be assessed, and a formula applied that brings the maintenance cost in line with this impact, based on the likelihood of an increase in required support resources stemming from the changes.

Since maintenance costs are usually worked out as a percentage of the initial contract value of developing the system, any increase can be represented as an indexed value that is derived from the complexity change, with the initial index set at 1.

Any future maintenance contract costs can then be recalculated with respect to the initial value of the system.

TRAINING

It is usual for the developer to provide some form of initial training before the client is able to fully realize the potential of the system that has been created for them. This is especially important in cases where the product is the result of the customization of an existing off-the-shelf commercial product.

This is due to the fact that, while the client may well have specified the features that they require in the end product, some of those may exist in the base product and therefore not appear exactly the same as those specified by the client. Thus, some training will be needed to help the user to become familiar with the product.

One final reason for offering good training is that knowledge is imparted from the developer to the client such that they can more easily help themselves, thus reducing the impact on help-desk resources in the future.

The best way to provide training to the client is by arriving on-site and performing the training in the operating environment. It will likely be a better use of resources, for simple economic reasons: it is more efficient for the trainer to travel to the user and train 10 users than for 10 users to travel for training to the developer's site.

Manuals should be produced that can be referred to when new staff members arrive, or when it becomes clear that the competence of existing, trained staff needs to be updated. They might also need to receive additional training if the core product or the customization changes.

Training Staff to Train Others

For this reason, it might be a good idea to always maintain contact with a single member of the client's team who is responsible for ensuring that the staff using the system remain competent to do so.

In addition, it might be beneficial to train a single staff member as if he will be giving training sessions in the future. In this way, he will be competent to give sessions on behalf of the developer, to new staff members, for example.

Finally, it might also reduce the impact on developer resources when the system is changed, since the staff trainer could amend the training package, indicating where new parts have been added and existing parts changed. With this information, he should be able to update the user base, without the developer becoming involved in additional training sessions.

SUMMARY

Delivery of the system is more than simply turning up at the client's door, wearing a suit and tie, and proffering a CD containing the application. It is a process that requires thorough planning and execution, and extends into what could be termed as the official maintenance period of the product development life cycle.

Usually a warranty period is specified, which gives the developer and client the chance to see how the product works in the field (like an extended piece of testing) before it passes into the phase covered by any maintenance contracts.

The two parts of the delivery can be viewed as:

- Physical Delivery
- Service Delivery

where the Service Delivery includes any on-site training and the warranty period covering the software itself.

Phased deliveries are also possible, where the client takes possession of a little piece of the system at a time, and can gain experience with it as the project progresses. This will save time at the end of the project, but is quite resource intensive, since each phase needs to be determined in advance, and then treated like a miniature delivery.

However, in cases where the system is an off-the-shelf one, with customizations made on behalf of the users, to their specifications, the phased delivery approach can be very useful in reducing the overall impact on the client in terms of resource management.

Whether the final product is designed to be sold to consumers, or whether it is a piece of programming that has been created solely with a particular client in mind, testing the delivery remains one aspect that can be overlooked. There are actually services available that will take the deliverable, install it on a variety of machines, and report the success of installation and removal under a variety of different circumstances.

The same service providers who provide these kinds of services will probably also be able to provide third-party user testing. While both of these services will demand a certain fee, it is probably worth the investment to make sure that the product delivered will actually work under most conditions.

For consumer packages, this can be considered part of the standard software development cycle, and needs to be budgeted for accordingly.

Part

III

Principles of Software Quality Control

The final part of this book equates software quality with end-user satisfaction, the "fitness for use" paradigm explored in the first two parts, and details how to deal with potential and real quality lapses.

Chapter 16, "Promoting Corporate Quality," indicates the various pointers that clients should be given in order to illustrate that the company embraces quality from start to end, and how to back it up with actions.

Chapter 17, "Testing Procedures," describes how testing and quality are related, and the ways in which certain processes can be put into place that ensure that the product is tested in a way that promotes quality, rather than as a simple contractual obligation.

A vital part of ensuring that quality is maintained is by establishing an efficient system whereby those responsible parties can be alerted that quality has not been achieved in their area. Chapter 18, "Feedback Techniques," deals with how communication lines can be set up so that any potential or actual quality deficiencies can be quickly addressed with minimal impact to the product and client.

In Chapter 19, "Client Satisfaction," we look at how the client can remain satisfied, even while a project seems to be failing, and what to do to ensure that they remain satisfied, if not with the specific product, then with the organization as a whole.

16 Promoting Corporate Quality

In This Chapter

- Introduction
- Projecting Quality
- Managing Quality
- Documenting Quality
- Summary

INTRODUCTION

When two providers are placed side by side, it is likely that the prospective client will gauge price and quality to ascertain which of the two offers the best ratio and is most likely to deliver a satisfactory end product within the specified time and budget presented by the client. Bearing in mind, as always, that the client understands that the chances of actually receiving a product that is exactly what they envisaged, within the time frame that they specified, and without overstepping the budgetary constraints that they put in place, are relatively slim.

It is important that the organization projects good quality, has well-documented quality control processes and procedures, and can back them up by actually carrying out these processes and procedures from the very beginning of their relationship with the client.

This chapter considers several areas:

- Communicating quality
- Managing quality
- Documenting quality

Each area contributes to the overall demonstration of quality to the client, and ensures that the quality is not merely documented but actually applied to the projects undertaken. Demonstrable quality will give an edge over the competition, and may even justify a higher price simply by virtue of being able to guarantee that the end result will be more reliable, deliver more functionality, and is more likely to be delivered on time and within the client's budget.

It is an unfortunate fact of the software development industry that many, many projects fall short of the client's expectations. The end result is either incomplete, inaccurately rendered, or takes longer and is more expensive to create than the client was initially led to believe.

Those software development companies that actually manage to deliver a product within the specified time, to specifications that were well understood by all parties, and is fault-free and reasonably inside the budgetary constraints will clearly have an advantage when it comes to competing for future projects.

This book aims at trying to achieve this Utopia by reducing reliance on logic programming, by reuse of well-tested components, and above all, by ensuring that the steps leading up to the design of the end product lead to documentation that reflects the product that the client actually needs.

There are various definitions of quality, but the one that stands out above the rest is that the product be fit for the use for which it was intended. This should be the guiding principle when trying to establish whether a product is "quality."

PROJECTING QUALITY

Corporate quality is dependent on everyone involved with carrying out tasks for the organization having pride in the services they are offering. It is a culture, not a directive; something that is important for management to understand from the beginning.

Employees cannot be ordered to carry out work that is of the highest possible quality; they need to want to do so because to do otherwise would lead to a result with which they would not be proud to have been involved.

Once the corporate quality culture is in place, it will show in everything that the organization does, from the first meeting with the client, to the delivery of the product, and in every piece of communication in between. As soon as there is a chink in the quality chain, it will show, and rather than projecting a high-quality image toward the client, they will be aware that there is a quality gap.

However, protecting the quality image to the extent that the client is not informed when there has been an error, or when the quality assurance might have failed, is not wise. Indeed, it is possible for the client to remain convinced that the organization follows practices of the highest quality, even while the project might appear to be failing; clients will appreciate honesty above all.

Communicating Quality

It is important not to confuse quality with image. Maintaining the corporate image means wearing a suit to the initial meeting with the client, if that is the image the company wants to project, or a more relaxed mode of dress, if the company takes a more relaxed approach to interpersonal relations.

In both cases, quality can still be communicated, since it is not really about the way people dress, but the way in which they present themselves in other ways. Quality is about having reliable processes to deal with every aspect of the software production process, with all employees confident that they deliver the highest quality service in their own specialization and contribute to the overall quality of the organization.

To be more precise, quality relies on having processes that govern the actions of the project members in everything they do. These processes must be predetermined, and deliver results that can be measured against the expected outcome of the process in question, with failures addressed in an appropriate manner.

Of course, the processes can be flexible, but they cannot be changed while they are being used. They cannot be adapted as a result of a failing, for example, while being followed.

Part of being able to communicate quality to the client is being able to demonstrate that these processes are in place, are being followed, and the results are being verified.

However, unless the client is informed that the various quality control procedures are in place, they will not know, and so the issue of quality control should be brought to the forefront of discussions from the very start. Leaving discussions of quality, quality assurance, and control until the discussion of the development process itself indicates that the organization might not have a quality culture, just a set of controls for the development process.

While this is better than having no controls at all, the software engineering paradigms presented in this book require that each stage of the process is governed by quality management. This requires having a process and expected result for that process for every stage of the software engineering cycle, from bid to maintenance.

Documentation

If each member of the organization is required to document his own quality level, there will be two side effects. The first is that there will be proof that the process has been followed, and that the results have been favorable. This is important, as projecting quality toward clients means that they have the right to ask for proof that the processes the developer claims to follow are indeed being followed.

The second side effect is that the employee is forced to be aware of what it means to follow practices that yield high-quality results, and what it takes to produce such results. If employees have also had a hand in creating the process, and set the levels by which their performance will be measured, they will have more faith in the process and hence are more likely to succeed.

The primary reason behind ensuring that documentation exists for each process and that the quality of each process is measured at regular intervals is so that everyone is aware of the standards to which the delivery of services is being held.

Documentation is also a tool for managing and communicating that level of quality. Even if the quality levels are lower than one might want, or expect, as long as they are constantly improving and can be proven to be so, clients can have a certain level of confidence that the organization will deliver on their promises.

Rewarding Quality

It is quite common for clients to offer bonuses for early completion of a project, and in some cases, this will mean that the savings made in time will be lost in paying these bonuses. Such bonuses have proven, in the past, to be a good motivator on a project-by-project basis.

However, they will not work over a long period. Management training courses are fond of pointing out that raising an employee's salary by a certain amount as a reward for good work has an effect that will likely wear off before the end of the second quarter of the year. Short of constantly raising salaries, it is not generally a good tool for ensuring productivity or quality.

However, providing other benefits, such as a company car, have a much longer lasting effect. In the case of a company car, every time the employee uses it, he will be reminded of his employer's generosity—especially if the employee has a certain amount of discretion in using it for private purposes.

Achieving high-quality results must be rewarded, not simply because that is what employees expect, but because if it is unexpected it is all the more pleasing. It should never become the norm, but if employees are aware that achieving a certain level of quality has a significant effect on the organization, they should be rewarded accordingly.

MANAGING QUALITY

Part of being sure that quality plays a role in the corporate culture is being able to effectively manage it. Managing quality requires that the quality assurance process itself is monitored, to the extent that it is governed by a process of its own, with its own set of benchmarks for success or failure.

This might seem a little extravagant, but there is a very real necessity to apply the same quality assurance mechanisms to the quality assurance process as is used to ensure that other processes are capable of delivering results of a certain quality.

To manage quality, we need to know several things:

- The process
- The benchmarks against which results of the process can be measured
- The current state of affairs

Each process needs a description, a set of stages with definite outcomes. For each stage there needs to be a set of metrics that govern the quality of the stage—this is distinct from the outcome, which we hope always to be favorable, but a process conducted at the highest level of quality may still fail (the bid process that results in the client choosing another supplier, for example).

We also need to be able to measure, at a given moment in time, where we are with respect to the quality plan. This is the stage at which management of the quality assurance process is the most crucial, and requires that a formal review take place.

Quality Reviews

The quality review is a step in the quality assurance process by which it is possible to assess the current state of affairs, and decide on corrective action if necessary. It has, itself, a series of steps, each of which must be carried out before the review can take place.

1. **Preparatory phase:** Collecting documents, selective reviewers.
2. **Review phase:** The meeting itself.
3. **Action phase:** The actions decided in the meeting are carried out.

This last phase will very likely extend into the start of the next phase of the project, being a part of the quality management process that will take shape over time. The preparatory phase involves setting up the review meeting and making sure that all the necessary information has been collected.

The review meeting needs to involve management, technical staff, human resources personnel, and project team members. The actual makeup of the review board will differ depending on the nature of the material being discussed in the review.

For example, if it is a technical review of a product design, there will likely be an emphasis on technical staff, and hence reviewed by a technically competent panel; possibly even members of other project teams, essentially the peers of those being reviewed. We will come to this approach later when we look at Quality Circles.

Deciding when to carry out quality reviews is also an important part of managing the quality process. Too many, and the project staff will feel that they spend most of their time sitting in review meetings; too few, and management will feel that they do not have a clear picture of the progress being made.

Reviews can also happen at different levels. Middle management, for example, might be content to know when each of the major phases is successfully negotiated:

- Tender and contract awarded
- Specification and design completed
- Development and testing completed
- Product accepted and paid for

If the software development process also includes staged delivery, such as a prototype, followed by deliveries offering increasing levels of functionality, the number of major phases might increase. A level of management above—at the department head and board level—might only require knowing when contracts are awarded and the ratio of failed projects to contracts won.

In the other direction, lower management and team leaders will need to carry out reviews more often, but usually as part of a more general review process. This means that the quality review meeting, while primarily aimed at measuring the success of the process against known benchmarks, will also serve to measure actual project progress as well.

This is not to say that it replaces project progress meetings, but that each project progress meeting will need to have a topic on the agenda relating to quality, and that an additional meeting will need to take place at the end of each phase that is dedicated to quality matters.

Quality Checklists

To measure the actual quality, we need to be able to break down the project status into a series of points with discrete answers. For example, there was a checklist when delivering a chapter of this book:

- Spell check performed?
- Verified format against guidelines?
- Proofread?
- Figures and tables included?

For each item, a certain level of quality has to be reached before the chapter can be delivered. Therefore, for example, the spell check must be performed, and all errors accounted for. Spell checkers, for example, tend to complain at source code, but these errors can be ignored. Therefore, while the spell check may fail, if it does so for an acceptable reason, we can proceed to the next item.

The document has to follow the guidelines, and, unlike a spell check, all violations have to be addressed by corrective measures. Proofreading will throw up some items that need to be corrected and others that can remain, so this is a point on the checklist that can "fail" but with qualifications. The same goes for the inclusion of figures and tables.

For software projects, it is impossible to give a standard checklist, but for an individual piece of code, the programmer might have something akin to the following:

- Are all designed functions implemented?
- Have all functions been tested with the standard test harness?
- Is the comment/code ratio acceptable?
- Does the code compile without errors or warnings?
- Has all temporary code been removed?

This checklist addresses issues relating to how the code has been written, how it performs with respect to known outcomes, and how well it satisfies the key points of software quality. It also serves as a way to measure progress, since the programmer is required to disclose how many functions have been implemented and whether they have been tested.

Total Quality Management

Part of the solution in adopting quality as a corporate culture is to follow a methodology that supports, and even promotes, quality awareness within the organization. Total Quality Management (TQM) addresses every aspect of quality and quality control, and seeks to provide the impetus that is required to ensure that the principles discussed in this chapter are applied throughout the organization.

Not surprisingly, the TQM approach began in Japan, where a tradition of honor and pride has been present in industry since the very beginning. In the West, it only began to become popular in the 1980s, 30 years after its conception in Japan. The difference in culture between Japan and the West means that companies find it difficult to implement TQM. John Stark Associates (*www.johnstark.com*) have identified that surveys by consulting firms indicate that the figure can be as low as 20 percent of organizations finding that they are able to implement TQM successfully.

TQM is, therefore, a cultural approach to quality in an organization, and, rather than being a dictum that is handed down from management also introduces the concepts at the product development and customer interface. This suits our approach to software engineering very well, as we can clearly identify the customer-facing part—the Liaison Center—and those who are involved in product development.

It is also a paradigm that fits with the general principles of software engineering, offering an emphasis on fast response times, with actions being carried out based on reliable research, and a sense of improvement, both in terms of quality and the quality process itself.

Something novel in TQM that needs to be applied to software engineering is that it is an approach driven entirely by the fact that it is oriented toward the customer. This is something that is reasonably novel in software engineering circles, not because companies do not want to please the customer, but because they did not have enough control over the processes involved to make communication between the company and the customer practical.

The introduction of a Liaison Center should go some way to address this, and provides a good starting point for TQM, which follows a client-first approach. This extends to every part of the relationship between the client and the organization. Traditional software engineering companies have assumed that it is sufficient to try to produce a product that matches the specifications, and that will make for a quality result. The TQM approach recognizes that this is only part of the solution, and that the customer-first approach needs to apply to every step of the process, as supported by the reporting lines in the Liaison Center discussed previously.

Another aspect of TQM that applies to this discussion is the dedication to improvement. If an organization is producing a product, this is an obvious path to higher quality, but a software engineering organization does not do this, and so the approach to take is slightly different.

In fact, the emphasis is on getting the design phase right, and before it, the requirements phase. "Continuous improvement" is a phrase that comes to light quite often in TQM discussions, and in this case, it needs to be applied to the process by which the product is created, since improving the product itself over time is not a viable option. The customer is not going to wait for several iterations until the programmers get it right, unless the rapid prototyping development paradigm is being used.

Even so, there will be a limit to the number of perfection iterations that the developer will be permitted by the customer to go through before they show dissatisfaction with the process. This will not yield a high-quality result.

Coupled with this is the notion of fast response—being able to react swiftly to the needs of the customer, or being able to repair a defect, should it arise, with a speed that ensures that the customer remains satisfied. We have pointed out in the course of our discussions about quality that a high-quality result does not necessarily mean that the software is perfect the first time around, but that the issues are addressed in a timely manner.

This is entirely in keeping with the TQM approach, and implies that a high-quality result can be achieved even if the product itself could be considered to be failing some of its own quality checks.

TQM must be seen as a driving force, and not just something that needs to be tolerated and reported on to keep management happy. It also needs constant drive so that it does not become forgotten once the novelty has worn off. The way to do this is to match the TQM implementation with the culture of the company, and to involve the employees with rewards and encouragement to ensure that the TQM mandate is carried through.

Quality Circles

Part of the TQM approach can be linked to a mechanism known as Quality Circles (QC), in which a small group of employees (never more than 12) meet together on a regular basis and discuss issues relating to quality. The meeting must be voluntary and not dictated by management, although a close eye will need to be kept on the scheduling of the meetings to try to spot trends and possible problems.

Part of the reason why the regularity of QC meetings will change from one set of employees to another is that there may not be sufficient impetus, or problems for them to meet. Put another way, if a meeting is to be held, but nobody has any issues to discuss, and the meeting is voluntary, then there is no reason to hold it.

Of course, this can be mistaken for either an unwillingness of the employees concerned to participate in the TQM scheme, or a sign that everything is as high in quality as it can be. The latter is easy to test for, and the former will prove more problematic, which is why we discuss the concept of documenting the quality process here.

The one thing that QC encourages is a change in attitude, and not just on the part of the management, but also the culture of the company as a whole. The employees participate through QC, which will encourage them to adopt an attitude of responsibility toward quality, and its improvement.

To adopt QC within a company that both embraces TQM and has followed the advice given in earlier parts of this book relating to the establishment of a Liaison Center, the company will need to superimpose a QC framework onto the organizational matrix.

The key QC roles are:

- Steering
- Coordination
- Facilitation
- Lead

The Steering function happens in high management, and will plan and direct the TQM program, as well as review the results passed up by the Coordination and Facilitation functions, which should be carried out by middle management, and will monitor the results of the QC scheme, based on reports from the Leaders of each QC.

Within each QC are a Lead function and a Participation function. Employees with no other role in quality control are the circle members, of which one needs to take the Lead role in reporting to the next level up.

Of course, the Coordination, Facilitation, and Steering functions will be carried out by more senior personnel, but this does not preclude them from also being QC members at their own level in the hierarchy. TQM and QC are, after all, part of the company culture.

DOCUMENTING QUALITY

One important part of being able to control the quality of the processes that are in place to ensure that the job is done properly—within the constraints of time, budget, and functionality—is having clearly documented procedures. This applies equally to the actual procedures used to create the software as to the procedures used to monitor the quality level and record defects.

The documentation serves two important purposes. The first is so that the quality assurance team can be sure that they have performed the correct steps in measuring the quality level, and resolved any issues in a way that is consistent with the processes that have been put in place to control it.

This in itself is important because it ensures that in rectifying a quality defect, another is not introduced in the system as a consequence of its resolution. This is linked to the second purpose of maintaining quality documentation—the audit trail.

If the developer wants to have the quality control processes that it uses, and the processes for software creation itself, validated by an external organization, then they need to be able to prove that the processes and their control have been correctly implemented. The only economical way to do this is to have the documents audited by a competent third party who will be able to validate that it is the case.

Process Description Documents

The first step in ensuring that the correct processes are followed and that each stage can be validated is to document them. Each process description document consists of a series of steps with verifiable conclusions that can be applied in every situation. If one has to create a set of process descriptions for each project, then this will unnecessarily waste resources.

However, it is perfectly acceptable, indeed likely, that for each type of product that is created there will be a different set of process descriptions. After all, designing and creating software will be different depending on the platform, interaction with other systems, and style of software. By *style*, we mean one, or a combination of:

- Single user
- Multiuser
- Real time
- Embedded
  - Interactive
  - Multimedia
  - Internet

These are the broad brushstrokes that we use to paint a picture of a piece of software—everyone knows that a single-user real-time interactive multimedia application is a game, and that the combination of skills and processes required to make it will be very different from a single-user Internet application, which might be, for example, a Web browser or FTP client.

Each of the styles needs, potentially, to have a process description made for it that can be integrated with other process descriptions to produce a guiding description of all the processes that need to take place to produce the product. Of course, this list, and hence this exercise, is not complete, but serves to illustrate that determining how something is to be made is a difficult proposition, but is also vital in being able to control and monitor that process.

Benchmark Reporting

The only way in which the processes that are described in the process documentation can be controlled is by measuring their success and comparing it against previous iterations of the same process. In this way, we can know if the quality is improving or getting substantially worse.

By way of a simple example, let us look at service-level agreements. A service-level agreement is a contract between a supplier and their customer that states that, over a given period of time, a certain metric has to be within a threshold of a value. Therefore, if the supplier is an Internet service provider (ISP), they might have a Service Level Agreement (SLA) with each client that states that their access will be available 95 percent of the time.

If the clients are home users, this will be acceptable, but if the user is a corporate client who requires the network being supplied to be available to generate revenue, then 95 percent is not going to be an acceptable availability rate.

Therefore, they might want to increase the SLA level, but to do so, they should not immediately boost it up to 100 percent, because the supplier will simply shrug, say that it cannot be done, and try to get out of the contract. Instead, they should benchmark actual performance, and offer an incentive to increase the SLA level by an appropriate amount.

This appropriate amount needs to be chosen so that it represents a substantially valuable increase from the point of view of the client, but the supplier is convinced that with a little work, they can achieve this new level of availability. This might be an extra 0.5 percent or an extra 0.1 percent. Either way, it should be agreed, and reported upon regularly, and when the new level is achieved, both parties can agree to raise it further.

This approach, which we call Benchmark Reporting, can also be applied in software quality management. We expect that there will be errors in products, but we can also expect that the number of reported errors per thousand lines of code (for example) can be reduced over time. It may never be zero, but at least we will be able to state a figure, know what the benchmark should be, and try to monitor our progress.

The key is to choose appropriate benchmark values that can be linked to the process that we have documented. Once we know what it is we want to measure, we can then benchmark the current state of affairs. Having done that, we can try to establish what level of that benchmark would represent an increase in quality, and try to put measures in place to achieve it.

Badges

We spoke earlier in the chapter about the possibility of having a qualified third party audit the quality assurance process and the processes used to create the end product, and the accepted industry standard for doing this was the ISO 9000 series of process quality control.

The International Standards Organization (ISO) has a whole set of quality guidelines that monitor the documented processes in many industries, from manufacturing to service industries, and taking IT in its stride. The standards are designed to be applicable across industries so they can be applied using the established process controls without changing them for each industry.

In this way, those implementing the ISO 9000 series can benefit from a set of standards that are exactly that—standard. Of course, there are guidelines that help each industry to implement the series in their own way, and plenty of advice on hand to help implement the controls that they prescribe. The certification process is expensive, but gives the right to display a specific badge that is taken by some as a seal of approval of the processes, if not the quality of the end product.

To be sure that the end product is of a sufficiently high quality, it needs to be looked at from the point of view of a third party. This third party should usually be the provider of the hardware on which the software is to run, or the operating system manufacturer.

If a company has received, for example, the "Designed for Windows XP" badge, we can be sure that the quality control processes that Microsoft has put in place have been applied, and that the product line is ready for that specific operating system and is guaranteed to deliver an experience that puts both companies in the best possible light.

Other industries have similar schemes. The video game console manufacturer Nintendo is renowned for the strictness of its quality control procedures, which means that any cartridge used in, say, its Game Boy range that has not been officially sanctioned by Nintendo effectively voids the warranty of the system. To add to that, only those that have been through the Nintendo quality control system may use the Nintendo badge on the packaging.

The theory is that without the badge, the consumer will not purchase the game. A similar theory pervades in the IT industry, especially among consulting firms. If they have not been certified for the processes they use, or the software, operating systems, or hardware that they profess to be proficient in, they may not be deemed worthy of using as a supplier.

Finally, training certificates are also a form of badge. Employees can be trained in a variety of different techniques and technologies, from programs such as the Microsoft Certified Solution Developer to hardware certifications from companies such as Cisco, IBM, and so forth.

Companies that can boast a large number of certified specialists will sometimes be given more leniency in their pricing than those that do not. This premium should compensate for the additional expense of training the employees so that they carry the certification.

SUMMARY

It is important to note that if an organization wants to compare favorably with the competition, it needs to be able to project an image of quality; and the only way to do this is to ensure that the whole organization has embraced the quality mantra.

Part of the problem is convincing people to have pride in their work so that they perpetuate the quality rather than stand in the way of it. Not caring either way will effectively mean that they are preventing maintenance and improvement of quality levels, so it is not an option.

The result is that processes and procedures can be put into place that yield evidence that high-quality standards have been respected, and thus the organization can apply for and hopefully win the right to display a badge that alerts the world to the fact that the company cares enough about the quality of their work to have the processes certified.

This does not necessarily mean that the software that is a result of the application of the certified processes will be of higher quality because of it. In fact, badges like the ISO 9000 series only show a commitment to using a high-quality process—it is the process itself and the quality controls that have been put in place that are being certified, not the software.

To be able to project an image of quality that holds within the IT industry it is therefore necessary to gain specific certification badges within those niches that the company operates. Therefore, if the company produces software primarily for the Microsoft platform, then they should ensure that their engineers are appropriately certified—the team leaders, if not all the project team members.

Involving the staff in the quality management of the company, which requires that they have pride in their work, will be much easier if they feel like they are achieving something as a result of their hard work—something for themselves—and training to receive awards and certifications both helps the employee and the quality level of the company. The bonus is that the employee will feel involved.

This involvement is a prerequisite for the application of systems such as TQM and Quality Circles, both of which require the active participation of all employees, and are also sanctioned for ISO certification. As long as no other process certifications exist that are applicable on a worldwide basis, these should become the lynchpin of any company's quality plan—whether they plan to certify or not.

All of this costs money, and while we can note that those companies that invest in quality certification and quality control may command higher prices on the open market than those that do not, it may not be necessary for niche players to certify their processes, instead ensuring that the technical certifications required to service the industry are obtained.

17 Testing Procedures

In This Chapter

- Introduction
- Consequences of Weak Testing
- Weak vs. Strong Testing
- Testing Implies Quality
- System Dependencies
- Testing vs. Certification
- Summary

INTRODUCTION

Testing is an integral part of ensuring that product quality is as high as it possibly can be. If the end product is poor, the faith in the developer that has been built up over the entire project can easily be destroyed if it is proven in their acceptance of the product that it has not been sufficiently tested.

In the preceding chapter, we looked at how corporate quality can be used in effectively negotiating and winning projects from other developers. Just talking about it is not good enough, however, and this chapter lays out ways in which testing can play a part in delivering on these promises.

Definitions of testing vary, as do definitions of quality; it is often best to look at testing as a mechanism by which it is possible to evaluate the product based on the satisfaction of two criteria:

- Robustness
- Correctness

The first of these is designed to ascertain whether the system is capable of reacting sensibly when encountering input data or operations that do not match a pre-selected set of operational conditions or fall within a set of predefined parameters. What amounts to a sensible reaction depends on the system, and one of the key definitions of system quality as "fitness for use."

This is linked to the correctness of the system, which determines whether the observed reactions are in tune with the requirements of the system as defined by the customer. For example, if the software is supposed to be monitoring a spacecraft, and it encounters a set of input conditions that seem to be at odds with the expected conditions, a sensible reaction will differ from a similar problem encountered by a less critical system.

Besides breaking down testing into these two areas, we can also layer testing over the traditional software development life cycle:

- Unit testing
- System testing
- Integration testing
- Acceptance testing

Unit testing takes place during development of each individual object; system testing is performed when these objects are glued together to create a part of the final system. Integration testing is used to ensure that the complete system performs correctly in its target environment, and Acceptance testing proves to the customer that the entire system satisfies the requirements captured in the opening phases of the project.

CONSEQUENCES OF WEAK TESTING

Two types of system test can be performed, which we will call weak and strong tests. A weak test is one that tests general cases in an attempt to show that the system works correctly in a vast majority of normal operating conditions. A strong test is one that ensures that the system is robust and can perform correctly in exception cases and in regular operations.

Weak Testing

Most of the testing that takes place during the development life cycle of the product will be weak testing. This is because there is simply not time to actually run through every single possible combination of data inputs and simulated operating conditions and verify that the results are correct.

This does not mean that a product that has been testing using weak testing methods is necessarily a lower quality product than one that has been subjected to strong tests designed to test cases that might not even logically be able to happen in the production environment. It does mean, however, that there is a higher possibility of an exception being caused that could lead to undesirable behavior, and that the developer needs to take the time to check that the anticipated impact risk is below an acceptable threshold.

Testing Process

To carry out weak testing, we need to be able to identify what constitutes normal operating conditions for the software product. This should be done with the client and can be based on simulated production data or live data that will be fed into the system to exercise it under a variety of different conditions.

The process is very simple:

1. Identify test data and cases.
2. Agree on expected results.
3. Test and report.

By agreeing on the expected results with the client, we introduce a separate process by which we can verify that the client and developer both have the same view of how the software is supposed to operate with reference to the problem that it is designed to solve.

Each of the test cases is then populated with a variety of data that has been agreed to provide an adequate representation of normal operating circumstances, including some exception cases that can occur during the system's interaction with regular users and other systems.

These test cases are then executed against the system, and the behavior recorded alongside other vital pieces of information such as time and date, as well as a reference to any screen shots or logs that can help the developer retrace the actions in the event of an error.

It is important that every test carried out is recorded. For example, during testing it is possible that a specific error occurs during the unrecorded execution of a test case, only to not occur when the tester decides to log a run that should, to all intents and purposes, be the same.

The developer will therefore find it very difficult to track down the origin of the problem, since they will have to work back from the only logged run, which turned out fine, without knowing whether there was actually an error or whether the tester misremembered the fact that an error occurred.

Once all the results are collected, they need to be reported to the client, who will then decide whether they are prepared to accept the product into the next stage of development, which could be either Integration or Acceptance testing.

It is the client's prerogative to decide whether any failed test cases have an impact that constitutes a high risk to their business, and therefore prevents the product from being prepared for release.

Data Sets and Scenarios

A Scenario is a set of operations that are designed to mimic behavior of the system in a production environment, and can produce results that are predictable and repeatable. Test Data refers to the information that is needed by the operations in order to provide information to the system for testing certain combinations of likely production data.

Of course, we can also use the Test Data and Scenarios to test exception cases as well, and a Scenario can be executed with more than one set of Test Data, depending on what we are trying to achieve. By and large, however, the idea is to gather information that can be used to validate a system for behavior that has already been tested to ensure that it is robust and correct.

When to Use Weak Testing

Weak testing is typically used in regression tests and integration tests. Acceptance testing can be seen as a form of weak testing, especially when it occurs later in the product life cycle; for example, to verify the correct operation of a change that has been implemented in a stable release of the software at the request of the client.

System and Integration Testing

The first time in the software development life cycle that we can envisage using weak testing is once the entire system has been unit tested and is ready to be put together and tested as a system.

It is a way of proving to the client that the system works as required, with reference to specific cases with which they will be familiar, rather then asking them to look at the unit testing results that suffer from being all-inclusive and hard to read.

The client may not realize that the combination of two unit tests that pass indicates that a specific feature of the system works as planned. However, a test case that touches all the various unit-tested functions and produces a result that they can associate with one or more of their requirements is far more useful to them.

Of course, these tests will still occur in isolation. It will be the system on its own that is tested, in an environment that is different from the end production environment. The next time weak testing is used is when the system needs to pass through integration testing, where it will be tested with the same set of test data and scenarios, and the results compared with the previous run during system testing.

Should an error be found, it is clear that it is in the way in which the system integrates to the production environment, hence highlighting a problem in:

- The Nonfunctional Requirements/Specifications
- The Design (interface and system borders)
- The Implementation
- The Test Cases

Under different circumstances, the impact of finding an error that results from a misunderstanding of the way in which the existing environment and the new one should integrate will vary from severe to "show stopping"; a "show stopper" being an error from which it is not possible to recover without calling a halt to the proceedings.

Regression Testing

During a regression test, we want to be sure that the system, as implemented, and probably as it is operating in the real world, will continue to operate in the same fashion under similar circumstances. It applies only to those areas of functionality that have not been altered between the full system test, release to the client, and the implementation of some new functionality. It can only test what has already been tested, and hopefully give the same results.

It is helpful to obtain some live data to perform regression testing, and to be able to compare the results of certain scenarios as laid out in the previous section. The live data should be easy to come by since it is assumed that the system that needs to be regression tested is actually in operation at the time the tests needs to be carried out.

Using live data will help to ensure that the weak testing is not hampered by lack of test cases; after all, we are only checking the software at a high level to see if the functionality that was offered before is still being offered. Since weak testing is not designed to verify that the inner workings are still 100 percent in line with expectations, a regression test without live data may work during testing and then immediately fail for some reason the moment it enters the production environment.

If an error is found during regression testing, the consequences can be very serious, since it can highlight that the change that has prompted the regression test, and that was not supposed to alter the existing behavior of the system, has. In turn, this means that the real source of the error is in one of the following:

- The original Functional Specification
- The original Design
- The original Implementation
- The original Test Cases

If the error is in the Test Cases or the Functional Specification, it will possibly prove equally expensive to fix. The supporting argument for this is that if the error is found to be in the Functional Specifications, they must be altered. In turn, the Design may also need to be changed, as well as the Implementation.

Then, the new software needs to be Regression and Acceptance tested. All of this will take many resources and be relatively expensive. If the Test Cases were the source of the error, and the software worked correctly under them, then exactly the same procedure needs to be gone through, with the new Test Cases in mind, because the original software was flawed anyway, or it would have failed in the testing phase.

Changing just the Design or Implementation will be considerably less expensive, although still more expensive than if the problem had been caught during Unit Testing or even earlier.

Acceptance Testing of New Features

For those parts that cannot be regression tested, it is assumed that new features have been added and need to be tested within the system. Of course, in line with the general theory of testing as a method to validate the software, the new features will already have passed through the unit-testing phase, and so will already have been validated using strong testing, as will the remainder of the system that has not changed.

The purpose of the Acceptance test is to be able to verify that the new features have been correctly implemented with reference to the rest of the system, and to be able to prove this to the client's satisfaction. There are several stages:

1. Decide on test data.
2. Set scenarios and expected results.
3. Perform and report.

The first two will need to be performed in conjunction with the client, prior to developing the new features. This is so that there is some kind of basic usage information that can be handed to the developers so that their unit testing will be more realistic. This is necessary because they will possibly not have been involved with the process that led to the development of the functions, and may not have been part of the original project team, and so may not have the required background.

In fact, since all communication will have been done via the Liaison Center, all information will be available to all parties, and this should not, therefore, present a problem. However, it is always a good idea to ensure that the developers of new functions have as much information as possible before they do their unit tests because it will save time in the end.

Next, using the test data information that has been agreed with the client but might not represent a complete set with respect to the actual tests that will take place, the test planners should construct a set of scenarios and expected results, in exactly the same way that was done for the initial system testing.

Then, these need to be discussed with the client to be sure that the expected results match the results that the functionality is supposed to include in the product. This is also a good way to validate that the correct change has been communicated to the developer, and that everyone agrees on the required behavior.

Finally, the tests need to be executed, and any deviation from the expected results noted and discussed with the client to see what the expected impact is. Should the impact represent an acceptable deviation, then this is noted and the software is released. If not, the source of the error must be found, which could be in a number of places:

- The change specification
- The change design
- The change implementation
- The original product

Since the purpose of testing is to highlight errors in any of the preceding cases, it is safe to assume that errors will be found that fall into any of the categories. It is also safe to assume that an error that is found during acceptance testing of the software, but is a consequence of the specification, design, or implementation of the original product will, on occasion, prove impossible to fix.

When we say "impossible," we mean "impossible within the resource constraints of the project." The argument is that since the error was found after implementation of the change, then since the specification and design of the change will have been performed with respect to the existing documents, the existing specification or design is at fault. Hence, it will prove very expensive to repair.

We can take it as a given that, under the circumstances, the requirements of the client have changed, since this is the reason for the change in the first place. Hence, we also have to alter the Requirements Definition and Specification document at the start of the process. This means that if we only find the error at the other end of the software change life cycle, it would almost certainly prove less expensive to create a workaround than to fix the error.

WEAK VS. STRONG TESTING

Weak testing is designed to ensure that the system can be validated with respect to normal operating conditions, as a way of demonstrating competence toward the client. Strong testing is a mechanism by which every possible aspect of the component under test is validated against a set of test data that is designed to take the system to its limits and see how it reacts. A well-designed system will have a contingency mode for dealing with possible errors and exceptions, and strong testing should catch individual instances but will not be able to identify system behavior defects.

The exception is when strong testing is carried out on a system that is in its Integration or System test phase. This is normally not practical, since the weight of test data and scenarios that is required means that it becomes an expensive procedure. However, in certain critical systems it is necessary to ensure that, at the very least, any possible operation that can be carried out by the user is catered for.

Imagine, for example, that we have a data entry subsystem that forms part of a larger application. If that component allows entry by the user of up to 100 characters of any in a specific range—for example, "A" to "Z" and the space character—then when we perform weak testing on the interface we might generate a set of 10 names and use them as the Test Data on the assumption that the keyboard to which the system is attached does not permit the user to enter any other characters.

However, when we perform strong testing of the component, we should ensure that we validate all cases, which means lengths and characters that are outside the ranges with which the system is expected to deal. We do this because, at some point, a new keyboard might be installed, and we want the component to be able to react in a logical way to changes that affect the way in which it operates.

When to Use Strong Testing

strong testing has a specific place in the software development life cycle—during the unit test phase and object test phase. In fact, we should use it for any testing that does not involve the system as a whole. We can never be sure what data will be fed into a component from any other component, but we must be sure that whatever happens, it is dealt with in a way that is logical and does not damage the rest of the system.

If the reader thinks back to Chapter 5, "Testing," we divided the entire testing philosophy into five separate phases: Unit, Module, System, Integration, and Acceptance Testing, each with its own set of test data designed to emphasize a functional area of the system. These were Memory, Data Type, Argument (or Parameter), and Logic. Table 5.1 shows the explicit relationships between these nine elements; phases on the one hand, and testing philosophy on the other.

We can further categorize the five testing types according to the nature of the testing that will take place—either strong or weak. Earlier, we noted that weak testing should be used in System, Integration, Acceptance, and Regression Testing, which is a special case that is generally only required after delivery and is dealt with as a separate project for the purposes of this book. It is worth pointing out that Chapter 5 tends to deal with the project in focus, and leaves out the discussion of regression testing, which we are now picking up.

strong testing, therefore, has its place in Unit, Module, and to some extent, System Testing, where we devise test cases to validate the Memory, Data Type, Argument, and Logic handling of the piece of the system we are implementing. There is, therefore, an overlap between Strong and weak testing in the area of System Testing. It is up to the project team to decide whether the product will require that the entire system is subjected to strong testing (rare), or whether we will limit this more rigorous approach to interconnected system parts only.

The Testing Chain

One might be tempted to insist that weak testing has no place in the Software Development Life Cycle, and that all elements must be tested as rigorously as possible at all stages, including the full system test. This approach will cause resource problems in the real world, where there is never going to be enough time to do everything. This is the guiding philosophy behind the testing procedures that we are proposing in this chapter—all the points that we made in Chapter 5 about devising test cases and so forth are valid for both Strong and weak testing, but the data sets will be more extensive when strong testing is performed.

That being said, the fact that we concentrate on Argument and Logic tests during the strong testing phases Module and System Testing indicates that we are not testing our components in isolation. In fact, we are ensuring that they will interact correctly with the rest of the system, and in doing so we also ensure that the risk of the entire system developing a fault due to inappropriate data entering the loop is minimized.

Hence, there is really no need to perform strong testing on the entire system, if we can be sure that the Unit, Module, and limited System Testing have all been performed according to the guidelines for quality control that the organization has established. There is a slight caveat, and that is where the testing (in an outsourced project, for example) is out of the control of the organization in question. In such cases, a special form of strong testing, known as Certification, needs to take place, and we cover this in a later section.

The Testing Chain ensures that each piece of the system that maps input data onto output data performs according to the logic defined in the Functional Specifications, and hence that when the entire system is put together, it will perform as expected.

In Figure 17.1, we have selected a part of a system that contains logic that maps some inputs onto two sets of outputs, via four functions, numbered 1, 2, 3, and 5. It is assumed that functions 4 and 6 provide some other service to the system as a whole, and fall outside the Limited System Boundary that we have determined is the part of the system on which we want to concentrate.

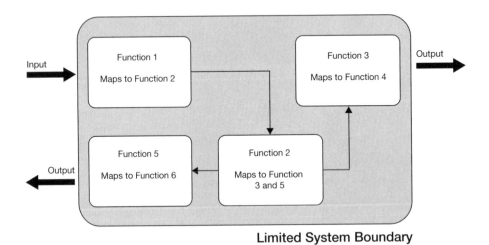

FIGURE 17.1 Test network.

Each of the functions will have been subjected to:

■ Unit Testing
■ Module Testing

Hence, we know that they behave in a manner that is both correct and robust; that is, they can correctly map their inputs to their outputs, and do not generate data that cannot be treated by the system, nor do they fail should they receive such data.

From a Test Network such as that in Figure 17.1, we can then construct a series of Test Chains:

■ Input → Function 1 → Function 2 → Function 5 → Output #1
■ Input → Function 1 → Function 2 → Function 3 → Output #2

We are therefore able to test that the Input maps correctly onto Outputs #1 and #2, using the Functional Specifications of the system as a guide. We can also verify robustness by varying the Input within the confines of a test data set that we have constructed for the purpose.

Since we have already been through strong testing phases for the individual functions, we can limit these cases to normal data and borderline cases, and still be sure that the system will function correctly under extreme cases.

The Testing Chain also ensures that one can pick up any part of the system and test it thoroughly and efficiently. Most of all, though, it also allows us to construct super chains, where the individually tested Limited Systems are chained together to create the final system. We can rely on the Testing Chain to deal with the robustness while the final system tests can deal with the behavior of the system as a whole.

TESTING IMPLIES QUALITY

We test so that we can be sure that the system works as intended when everything is running according to the expected behavior of the entire system and the environment into which it is placed. We also test so that we can be sure that the system is not going to fail completely if something unexpected occurs.

One of the common myths associated with quality and software engineering is that a quality product is one that has been tested thoroughly. As we shall see, quality relates to testing in several different ways. For example, a product that has been extensively tested using strong testing may well be robust, but there is no guarantee of correctness of behavior if there has been a misunderstanding in the fundamental way in which it operates.

On the other hand, it is difficult, if not impossible, to create a product that is of high quality without performing any testing, and all high-quality products will have been tested extensively. Subsequent iterations of the product, should it be subject to changing requirements down the line, will be much easier to implement if the whole quality matrix has been filled. If the product has not been properly tested, it may be difficult to be sure that the changes will not cause unexpected behavior in the modified system.

Correctness

A product can be said to be correct if it behaves within the constraints laid out in its design with reference to the expected conditions under which it will be operating. This means that a correct piece of software can still fail to satisfy the client, but it will have been verified to be correct with respect to the design. If the design is flawed, then even a correct product will not be of high quality.

However, if the product is not correct, it will stand very little chance of being acceptable to the client; it will not be of sufficient quality. Validating correctness requires that each of the design areas can be verified with reference to a set of test cases and data.

Correctness is established by using weak testing, since we do not need to cover every possible combination of data and operation. However, in the process of validating the correctness of the software product against the design, we may well find that we have been able to cover as many cases and combinations as will occur in the normal operating environment.

Of course, it is not only the software itself that can be tested for correctness. In fact, the Design should also be tested for correctness against the Specifications. How this is performed will depend on the format that has been chosen for the Design, but will probably have to be performed manually.

If we go back still further, we can actually test the Specifications against the Requirements too, but at this stage, we are going to be performing a set of tests that are qualitative rather than quantitative. By this we mean that, before the system becomes concrete it is impossible to get empirical proof that it is operating according to the definition to which we are trying to compare it.

Qualitative evaluations of the correctness of the system tend to take longer to perform, and yield results that might be accurate but are not proof that the system will work. They will only prove that the author of the document has understood the principles well enough to commit them to paper. The evaluation of whether this is the case will need to be performed by the client by open discussion.

Robustness

The principle of testing for robustness is to be sure that the system is solid, and that unexpected events or input data will not cause it to perform in a way that is unacceptable. The definition of unacceptable behavior will differ from system to system; in some cases, what constitutes acceptable behavior in one case will not be acceptable in another.

In fact, dealing with exception cases is an example of a testing topic that can fall either into correctness or robustness. If there is a specifically defined mode of behavior that is designed to deal with exception cases, then this falls into the category of validating the system for correctness of behavior.

If the system has no mechanism for dealing with exception cases of a certain kind, this can be classified as testing for robustness. In other words, in the absence of specific error handling routines that provide behavior that indicates that an error has occurred, we need to at least be sure that the system will continue to operate, even if it does not inform the user or client system that it has failed.

The worst case is one where the system simply ignores the data and continues to operate without treating it as best it can. This will usually have the effect of snowballing until a part of the system makes a decision that will cause the process to terminate. As long as nothing occurs that might damage the integrity of the system, and any associated data, this behavior is acceptable, and the system can be said to be robust.

However, should the system fail—that is, cause damage to its integrity, terminate processing completely, or fail to produce the required result—then the system is not robust, and this will need to be addressed. Whether this requires that a part of the system is redeveloped, or part of the specifications altered, or simply a note made in the User Guide is a question that only the client can answer with any degree of certainty.

Correct Behavior and Fitness for Use

We noted earlier that a failure in either Correctness or Robustness does not necessarily imply a product that is not of the quality expected by the client. While it is true that it can be said to be of a lower quality than if the errors did not exist, part of the formal definition of quality includes a phrase such as "fitness for use for a given purpose."

In fact, this is often part of the standard disclaimer that can be found attached to the license of commercially available software; except that in such cases, it is usually in the negative—no provision is made for the guarantee that the product is fit for a given use or purpose.

When software is being created specifically for a client, we need to be able to claim that the result will be fit for the use for which it was intended, and this is part of the reasoning behind trying to ensure that everyone has understood, via the Requirements Definition, Specification, and subsequent Functional Specification, what the delivered system is supposed to be fit to do.

Thus, fitness for use is a contractual obligation. If the product cannot be used, then the developer will not be paid, or at least, they will not be paid beyond the money that they will already have asked for by way of advance.

Should the system violate some of the criteria set aside to determine if it is correct, it is up to the client to decide whether it is still fit to be used. We would like the system to be 100-percent correct, as would the client, but everybody understands that, due to the nature of software, this may not be possible. In such cases, there needs to be a mechanism to accept the software "as is," on the assumption that it is fit for use.

Determining Fitness for Use

While, to a certain extent, Correctness and Robustness are both empirically testable in that they rely on sets of test data that may input data or behavior onto output data or behavior, Fitness for Use is a different prospect altogether. It is a good idea to decide what constitutes Fitness for Use as part of the Acceptance Testing cycle.

The danger is, in deciding on the Fitness for Use conditions earlier in the SDLC, that the developer will look on that as the real test, and not the actual Correctness and Robustness test data sets. Thus, while the product will be Fit for Use, it may not be of the highest quality possible.

To establish what might be constituted as adequate tests for Fitness for Use, we need to look at the functional areas of the product and decide whether it delivers on them in a way that satisfies the client. This often makes it look like a second-best option, a fallback position in case the software does not function correctly, but this is not the way in which it should be viewed. The test for Fitness for Use must be carried out regardless of whether the system has been deigned Correct and Robust, or not, because it is the only way that the client can be sure that it is applicable in solving the problem they identified.

SYSTEM DEPENDENCIES

In our previous discussion of Test Chains, we noted that Limited System testing is based on mapping a series of inputs onto outputs, assuming that the same has been done for all the components in the chain, and measuring the result against known, or predetermined, result sets. The system as a whole can be viewed in the same way.

Many of these inputs and outputs, with reference to the system as a whole, will be linked to dependencies on other systems. As such, poor overall system quality can sometimes be a consequence of other system or peripheral failures. This is a sign that those systems or peripherals have not been tested according to the same principles to which the system under development has been subject.

Worse still, they have deficiencies that were not foreseen by the development team, and therefore only found during the testing phase. Usually, when this occurs, we decide that the system must be redeveloped. At the very least, the failing component needs to be examined to ensure that the changes that need to be made to correct the problem are correctly passed back to the specifications of the system.

In our discussions about software development in this book, we have often concluded that if this needs to be done at such a late stage, it will probably cost too much to go through all the various processes, and therefore the error will be logged, but left unrepaired.

However, we need to have a mechanism by which we can deal with this when it happens in the domain of a system dependency, because it is not part of the system being developed, and although the same rule holds true—it will be more expensive to fix at this stage than if it was found in the opening phases of development—the chances are that a system dependency failing will produce a net result that violates our Fitness for Use principles.

Prevention and Cure

Of course, the best way to deal with possible quality degradation as a result of system dependency flaws is to isolate problems as soon as possible. System dependencies can be broken down into several categories, and testing procedures need to be established that can deal with each of the categories effectively. These categories are:

- Internal tools/drivers/components
- Third-party Open Source components
- Third-party Closed Source components
- External hardware devices
- External software systems

The way in which problems stem from dependencies on the preceding system types varies by category. In each case, however, the goals are the same: to identify errors or failings, correct them if possible, sandbox them if not, and protect the system against such errors as may affect Fitness for Use.

Internal Tools, Drivers, and Components

This category of System Dependencies relates to anything that has been developed to provide extended functionality beyond the scope of the Functional Specifications to achieve one or more of the Requirements Definitions but that does not form a part of the system being tested.

It also applies to those specific pieces of code, such as device drivers and libraries of additional functionality, that are required by the system, have not been developed as part of the system, but for which the developer has a case history.

This applies to pieces of reused code, taken from the Object Repository, whether they have been extended or not, other code that has been developed for other projects but has a relevancy to this project under scrutiny, and specific drivers that are needed because there is a specific piece of hardware that has no standard drivers available.

Since the code comes from within the developer's team of software engineers, it is easy to be sure that the code on which the system is dependent has been tested. It is also easy to know to what extent and success the code has been tested. The decision can then be made as to how to proceed.

If there are no defects, and the code has been tested extensively enough that the developer is sure that there will be no defects when it is integrated with the system being developed, then all is well, and the components can be used. Should they subsequently fail, they have to be treated as little projects of their own that have defects that need to be addressed in isolation. This minimizes the overall impact on the system, made possible by a combination of strong and weak testing within the testing philosophy, and by relying on Testing Chains to reduce the necessity to constantly re-test objects unnecessarily.

If previous testing has not been sufficient to make a decision as to whether the components are fit for use, the developer needs to do several things. The first item on the list is to find the responsible engineers and remind them of the importance of strong testing in Unit and Module test phases.

The second is to complete the remaining tests, which will ensure that there is a complete test case history for the components under scrutiny. Once this has been performed, the developer should be in a position to decide whether these components are fit for use. If they are, they can be used, with the caveats mentioned previously relating to any subsequent defects.

If, in either case, the components are deemed to be of insufficient quality for inclusion in the project, the developer has either to address the defects, create a component that is more Robust, Correct, or both. It can then be passed back into the Object Repository with the knowledge that the collection has become of higher quality.

Finally, there is the possibility that the developer decides that the components are beyond redemption. In this case, they will develop a new component, and return it, tested, to the Object Repository, again knowing that the overall quality has increased and the range of available components has also improved.

Third-Party Open Source Components

These components are pieces of code that have been assimilated into the Object Repository or Component Gallery, and are Open Source. The exact meaning of Open Source is discussed in Chapters 11, 12, and 14, where we look at the implementation side of software engineering. Assuming that the contractual obligations for the code remain fulfilled, the organization is free to use, develop, and maintain the Open Source components, and this includes multiple reuse across projects.

Subsequently, we can deal with them in the same way as with the Internal Tools, Drivers, and Components listed previously. There might be additional feedback requirements, depending on the nature of the Open Source license (GPL, LGPL, Public Domain, etc.) under which the code was released. These extend to ensuring that any improvements are handed back to the community such that they can enrich their own collection.

Third-Party Closed Source Components

Of course, there will also be occasions where the Component Gallery contains libraries, tools, or drivers that are Closed Source. In other words, the source code is not available to the developer; all they have are the rights to use the object or executable code.

Should this not be the case—if the developer has purchased the rights to the source code as well—then they can be treated in the same manner as Open Source components, as detailed earlier.

In other cases, while the components can be tested, they cannot be altered should a defect make itself known. Remembering that we are dealing with actual defects, and not just cases where the functionality does not meet our requirements, we need to ensure that the noncompliance is noted in the Component Gallery as a first priority.

This will usually cause a chain of events in which the Liaison Center, along with the Librarian or whomever is nominally in charge of the Component Gallery, informs the original developer, and follows up any solutions that they offer. Should the original developer offer a workaround prior to adjusting the code, this needs to be communicated to the engineer who brought up the defect in the first place.

Next, there is a decision to be made. If the component is the only one of its kind (specialist applications and hardware devices, for example), it is likely that the developer will continue to use it regardless of its specific defects. This is based on the assumption that the code is fit for use, with caveats.

The developer will need to try to sandbox the component so that the defective behavior either goes unnoticed to the system being developed, or an appropriate behavioral model is adopted. Sandboxing usually demands that some code be written that forms a buffer between the component and the system under development.

Should there be other options, the developer will need to evaluate them, and see whether they should be used in place of the defective component. If they need to be acquired from an external source, the efficiency and cost effectiveness of doing so needs to be evaluated alongside. The developer needs to bear in mind that in testing the new component, additional defects might be found.

External Devices or Systems

Much of what we have so far said is applicable to those external systems or devices upon which the software system under development needs to rely. However, testing them should begin much earlier in the SDLC than for other components, since they are likely to be part of the Requirements Specification, especially in cases where the system needs to interface with nonstandard or exotic hardware and software systems.

Consequently, it is more likely that the developer will choose to sandbox the components directly, by placing some code between the misbehaving device or system and the system under development such that the effects are minimized.

TESTING VS. CERTIFICATION

The final topic in the discussion of Testing Procedures and how they relate to overall software quality control is a little different from the others. In fact, Certification is used in cases where we do not know exactly what testing might have been done—it is a service that is in place to ensure Correctness and Robustness, usually applied to outsourced projects or independently developed systems conforming to a set of specifications.

To perform Certification of a system, we must:

- Have a set of Functional Specifications
- Determine sets of Test Cases for Robustness
- Identify Scenario-based Test Cases to verify Correctness

It is also usual that Certification is applied in cases where we own the Functional Specifications, have given over the development of the system to a third party, and have trusted them to perform the testing part. Upon delivery of the final software, we must then Certify that it conforms to the Specifications.

In the context of a software engineering house, this may not seem relevant, until one remembers that we could choose to outsource component development to ease the burden of delivery to the client. In such cases, Certification becomes a viable way to verify that the delivered product conforms to the Functional Specifications and will perform under exception conditions.

Why Certification?

Certification is distinct from testing in that it simply looks at the system as a black box into which we can feed information, and from which we expect some kind of behavior. While we use Certification to validate certain border exception cases with regard to logic, we will only test for robustness at a data validation level.

In other words, while we will try to ensure that the system does not produce data that is inconsistent with the environment in which it is operating, and does not fail when confronted with data that it was not expecting, we will not try to trick the system by exposing defects in its logic except when those cases might arise under normal operating circumstances.

The other side to Certification is that, unlike testing, it is assumed that only one set of tests should need to be executed. In other words, we test to expose defects in the system or component, but Certification exists to ensure that the testing has been performed correctly, and we do not expect to find any defects remaining.

Taking this one step further, we can also state that if Certification fails, it is most likely to be due to a misunderstanding with respect to the Functional Specifications, rather than insufficient testing. We would hope that the developer had placed sufficient emphasis on the development and testing process that this would be the case.

Certification Test Cases

Looking again at Table 5.1, note that Certification is not covered by the various phases and test types mentioned in it. Certification is, in fact, rather a mixture of Data Type and Logic validation. Therefore, we want to test for robustness and correctness, but in a way that does not preclude any knowledge of the workings of the system.

In the same way that the Definition and Specification of the system make little or no reference to how the system should be implemented, the Certification of the system should verify that it operates correctly in a similar manner. Therefore, only the user interface that is exposed by the system may be used to introduce data into the system, which effectively reduces the amount of Robustness validation we can do.

Therefore, in designing the Certification Test Cases, we need to be aware that we are only able to test the Robustness of the system by testing the Correctness. By way of example, we could design a system that needs to receive a number via the graphical user interface.

If the system has been created in such a way that the data entry field will only accept numerical data, then we are confronted with two issues:

- We cannot test alphabetical data with the system.
- We do not need to test data entry beyond simple, reasonable test cases.

From a testing point of view, this is not acceptable, except (again with reference to Table 5.1) when performing Acceptance or Integration Testing. In fact, Certification replaces Acceptance Testing in an SDLC that has been outsourced.

SUMMARY

In this chapter, we looked at the various testing types that have to be put into place, and presented loose descriptions of the kinds of procedures that need to be implemented, but without giving precise instructions. The reason why we were a little vague about exactly how testing procedures should be implemented is that they are dependent on the organization's way of working.

However, there is merit in providing a series of steps that need to be followed:

1. Create test cases.
2. Execute test cases.
3. Report results.

Exactly what has to be done in each phase depends on whether we are performing strong testing, weak testing, or Certification. Chapter 5 provides information on how to create the test cases and report on the results; the discussions here relate to when and why the testing needs to take place.

In line with the continuing emphasis on reuse and component-ware, we should also note that each product that is created should be tested with the emphasis on testing new code from a virgin build, and basing the test conclusions as far as possible on the results of tests applied to those components that are being reused.

Where this might become more difficult to justify is where a component is being used for a purpose for which it was not originally intended. In such cases, any new code must be tested from a virgin build—it assumes that changes have been made and that the sources are available.

18 Feedback Techniques

In This Chapter

- Introduction
- Reporting Line
- Central Communication—The Liaison Center Revisited
- Supporting the Reporting Process
- Summary

INTRODUCTION

Part of ensuring that high quality is delivered at each stage of the product development life cycle is being able to ascertain where lower than expected quality has been delivered, and being in a position to put into effect changes that result in an increase in quality. Feedback is important in being able to identify and act on cases of lesser quality, and there are specific ways in which this can be achieved.

Not only should we work toward being able to identify actual deficiencies, but also cases where quality may potentially be lower than the client has come to expect. Quality risk assessment is a reasonably new concept in software engineering, if only because it costs money to perform but does not actually bring any income on its own.

Risk assessment in general works on the basis of fear of failure, estimating the costs associated with a particular part of the system failing, and assuming that, at some point, it will fail. Quality risk assessment works on the basis that in order to achieve a higher level of quality, we need to invest in avoiding cases where we are unable to gauge the likely level of quality that will be delivered.

This chapter deals with the various forms of communication that need to be built up in order to be in a position to determine what the quality outcome of a particular exercise is likely to be, as well as using models to try to anticipate failures in the quality process so they can be avoided—at a cost.

REPORTING LINE

Testing usually occurs at the end of the Software Development Life Cycle, either of the system as a whole or of the component that is under test. Table 5.1 indicates the various phases in which testing takes place—each phase has its own area of responsibility regarding what needs to be tested and to what extent.

If a defect is found, the consequence is that the nature of the defect needs to follow the reporting line back to the design phase (in extreme cases) so it can be corrected in the most efficient manner. This means that we need to know which member of the team was responsible for the design and development of the component so that he can be the one to repair the defect.

The assumption is that the problem is found in a timely fashion, and that the member of staff responsible for the defect is both still working in the organization and on the same project, although not necessarily on the same component. The chances of this being the case decrease the further into the SDLC the software progresses.

Documenting the Reporting Line

Depending on the size of the organization and of the project, the team responsible for ushering each piece through the SDLC will be different at each point. That is to say, the staff doing the Acceptance Testing may not have access to the source code written by the original developers. The staff performing a Module test might not be the same developers as those who created the units that they are gluing together, and so on.

In addition, those responsible for the Specifications might not be the same as those who have found the error. It may be the case that the component needs to be passed back to the Definition stage, in which case there needs to be some way to ensure that the team can be assembled to repair the defect in question.

Of course, this is entirely dependent on the seriousness of the problem, and the point during the SDLC at which the defect was found. Each component needs to have a textual definition that accompanies the source code, then the executable or library code, from Specification through to Acceptance.

There should only ever be one copy of this document in circulation, and it provides the reference to the entire Reporting Line. At a minimum, it needs to list the names of the staff who worked on the component, in the following areas:

- Specification
- Design
- Implementation
- Integration

In addition, each entry needs to have a testing reference. We have not dealt with explicit references before in our discussions of testing, but it is the Reporting Line document that needs to contain the explicit reference to the documents that describe the tests that took place, and who signed off the results.

Each testing section needs to refer to the standard tests that were performed, if any, and the specific tests that were performed to ascertain that the logic had been correctly implemented with reference to the Specifications and Design. For example, the organization should have a standard set of tests to deal with validation of common data types:

- Integer and floating-point numbers
- Strings
- Memory blocks and pointers

These will be used in the Unit Tests and possibly Module Tests to ensure that the interfaces to the functions have been correctly implemented. By the time we are ready to run System and Integration Tests, we will be using scenarios designed to validate the operation of the system using test data created especially for it.

All of the tests need to be referenced in the Reporting Line document, with the intention that exactly what has and has not been tested is entirely transparent. This should be seen as an effort to help in resolving the problem as quickly as possible, when it occurs, rather than an attempt to point the finger at specific staff members when they accidentally neglect to either test or report effectively. The concept of the Reporting Line needs to be integrated with the Quality Assurance philosophy of the organization. For example, in the opening chapter in this discussion on Quality

Control, we spoke about Quality Circles and Total Quality Management as a way of trying to gain some level of measuring and controlling quality in the processes used.

Thus, the Reporting Line document might become a part of the Quality Circle discussion (a voluntary meeting), and will certainly be part of the whole Total Quality Management cycle. Since the idea is to instill corporate pride in the level of quality delivered, it is clear that far from being a document that the staff will be afraid of, it is in fact a document that is there to help them, and they should feel proud about being a part of it.

The Reporting Line Document

While we discussed the minimum structure of this document earlier, each section will contain slightly different information depending in which stage of the SDLC it was reported. The information will also reference other documents that are used in each of the SDLC phases. It is almost a nominative summary of the entire process from conception through to delivery that monitors the progress of the software or software component until it is fit for use.

In fact, there will be many Reporting Line Documents, some of which will only come into being as a result of a specific decision. For example, during the Functional Specification phase of the SDLC, a number of individual functions will be isolated. Using the Object-Oriented Design paradigm, these will then be turned into objects ready for implementation.

Since each object is isolated for the development process, ready to be glued together during integration (whether it be a object created specifically for this project, or something from the Component Gallery or Object Repository), a number of Reporting Line Documents will have to be created to follow these objects through to System Integration, where the results will be merged back into one document.

Thus, for a given project, the documentation collection will look something like Figure 18.1, where some Reporting Line Documents will have a lifespan that ends with the successful testing of the respective components, but are referenced in the final Reporting Line Document.

Therefore, following Figure 18.1, the original Specification becomes the subject for the Analysis, which leads to a number of documents, each referring to a specific function in the design. These then become multipart documents, each referring to a specific component that makes up the functional area to which it refers. Once tested, they leave the Implementation Phase to be merged back into a single, multipart document that provides the final summation of the Reporting Line.

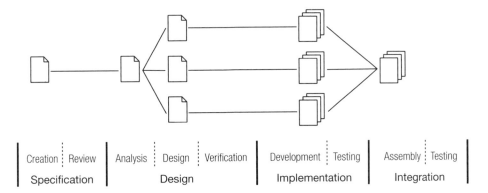

FIGURE 18.1 Following the Reporting Line Documents.

It is subsequently possible to go back through the documents and find out who was responsible for which part of the error that is under analysis at any given time. There should be references in the source code to the Reporting Line Document so that it becomes a simple matter to follow the chain back to the initial Specification.

Specification

The Reporting Line Document for this phase needs to show the team responsible for the creation of the Specification, and what sources have made up the document (Requirements Specification and/or Definition). There also needs to be a reference to any Quality Circle meetings that discussed the Specification, and a reference to the sign-off document that the client should have submitted.

The sign-off document, possibly part of the contract, is proof that the clients have reviewed the Specification, possibly in conjunction with the project team, and agrees that it provides the functionality they require to solve the business problem they have isolated.

Design

While the Reporting Line Document starts as a single document, it is necessary to create several to follow the different functional areas of the project. This is because there will probably be multiple teams, and to avoid problems with updating documents and central storage it makes more sense to spawn several documents than to create a single multipart document that everyone might need to update.

During the design Verification stage, each of the documents needs to reference the specific sign-off documents and the kinds of verification that took place. For more information about this aspect, the readers should turn their attention to Chapters 8, 9, and 10, where details are given about proving designs based on specifications.

Implementation

Between the Design and Implementation phases, the Reporting Line Documents that have been created to monitor the progress of the individual functional areas become multipart documents, with one part per functional object. They can be grouped in this way because the team that handles the objects for one functional area remains the same.

In cases where there are multiple teams handling a single functional area, or where an object in that functional area is being handled by another team, it is acceptable to create another Reporting Line Document, as long as references to it exist in the parent, and the parent does not continue to receive entries for that object.

During testing, it is important that the test documents and test team are referenced explicitly, as well as the results. The Reporting Line for each of the objects is now complete, and only the references to these documents will remain beyond the end of the Implementation Phase.

Integration

Finally, the multipart documents that make up the entire Reporting Line reference for the project need to be collated, referenced, and placed in a single, multipart document. Then, references are made to the documentation that specifies how the objects are to be glued together, and the team responsible.

Finally, the testing team and results need to be noted for each of the separate test items that were prepared to ensure that the Reporting Line can be completed, as it was in the Implementation phase.

The result is a document that contains many references to previous documents, and a brief overview of the development history. This ensures that, when an error is found, we can trace the origin of the code and every staff member who has been involved with it.

Again, this is not so much to assign blame, but more to try to ensure that the solution is found as efficiently as possible.

CENTRAL COMMUNICATION—THE LIAISON CENTER REVISITED

It is not possible to maintain efficient communication of feedback documentation without a central point of contact, and the Liaison Center provides that central point. The issue is that there are a large number of documents that need to be maintained and exchanged, and if we allow a network of documentation relationships to build up, there will be cases where a specific set of documents can no longer be found.

The Liaison Center is also important in managing the overall corporate quality ethic, and as a central point in managing relations between projects, teams, and clients. Not to mention the role that it will play in setting up various meetings and reviews to measure the success of the organization in delivering on their promises.

It is also a good idea that the Liaison Center carries out specific reviews of quality issues because they can be seen as a disinterested third party. This is only possible in organizations that have the resources to ensure that the Liaison Center does not comprise staff who also have a role in other areas of the company.

Quality Management

Before we look at how we can measure and monitor the evolution of quality metrics, we need to explain why it is important. More importantly, we need to look at why, in the context of Software Engineering, it is a good idea to be able to measure quality empirically, and what we can do with that information.

Part of the problem is that computer applications, as we previously mentioned, are not like other products. They are intangible, and the only way in which we can be sure that they have been correctly put together is through their use. They are also transient—they can be changed much more easily than other, concrete products.

This leads to two important facts of software engineering. One, a change is easy to make, but often far reaching—it is quite possible to fix one aspect only to break another—and the result of the change is not always immediately apparent. Two, a piece of software might almost work, even when there are aspects that are clearly broken, and still be fit for the use for which it was intended, even when the defects are more than just superficial.

The Liaison Center is the point at which all the quality lines converge. Consequently, it is also responsible for handling Change Management, Total Quality Management, and Quality Circles. After all, it is in touch, or should be, with all of the different departments and teams within the organization, whether it is a small development house with 10 people, or a 1,000-strong industrial software development company.

Change Management

One of the principle ways in which we can measure the success of a product is through monitoring the Change Management. For example, we might have an unusually high number of Problem Reports that are passed from the System Test team back, via the Liaison Center, to the Development team. This might, depending on other metrics that we will discuss later, indicate a problem with the way in which the development has been done by that team that will have repercussions when the product hits the field.

However, Change Requests passed back from the client to the Liaison Center, during the Acceptance Test process, might indicate that there are problems with the initial understanding of the client's request. This also has to be dealt with in a certain manner, taking into account other metrics.

Change Management is typically one of the areas of software engineering that is the cause of many quality problems. The fact is that if the development teams are allowed to run their own change management procedures, this will eventually have a negative impact on the quality of the product. The way to avoid this is to have the Change Management process managed centrally.

Total Quality Management

The Liaison Center also fits into the TQM paradigm perfectly, since the emphasis of TQM is on a customer-driven approach to providing goods or services. In this case, the Liaison Center treats everyone as either a client or a supplier, whether internal or external, which is perfectly in keeping with the TQM mindset.

To embrace TQM, we need to be sure that we can generate actions that are geared toward improving the quality of the organization's processes and products. To do this, we need to base our evaluations on factual information, which in turn implies that only a single central point of contact will have access to all the relevant information.

By establishing contact between customers and suppliers via the Liaison Center, we can also be sure that any defects that occur in the process are dealt with before the baton is passed to the next stage in the SDLC. This is important because, as we have noted many times, the cost of repairing a defect increases rapidly if we need to take it back more than one stage in the process. This would not serve TQM, and so we would rather pass all communication through the Liaison Center and deal with defects as they are located.

Quality Circles

Finally, the Liaison Center is implicated in Quality Circle administration, monitoring, and even as a participant. Since the Liaison Center will probably provide the information to the nominal leader of each QC initiative, it makes sense that they should be on hand to monitor how this information is being used, so that they can improve their own processes.

Of course, the question then becomes—who is monitoring the Liaison Center? The answer is that TQM and QC philosophies dictate that this should not be necessary since we are embracing a quality culture.

The real answer is that the management team, also in tune with the TQM mindset, is responsible for making sure that quality is a top-down affair, and that each level below them manages its own quality appropriately.

Hence, the Liaison Center will conduct its own QC, and the results of this should be passed up to management so they can verify that the culture has taken root in a manner of which they approve.

Quality Measurement

It is not possible to engage in Quality Management, as we detailed previously, without some way of measuring progress. In Chapter 16, "Promoting Corporate Quality," we stated that part of the overall philosophy that needs to drive a corporate quality program is in improving, little by little, until a certain goal is reached.

Measurement fulfills two roles. On the one hand, it allows us to see the current state of play, and how we can improve it in the future. On the other, it provides us with a way to monitor how the quality improvement process is developing over time, which may offer clues as to how it can be better implemented to achieve more rapid change.

Measuring Quality

To accurately measure quality, with a view to actually verifying that the situation is improving, it is necessary to try to establish indexes that reflect the nature of the defects. In fact, it is very difficult to measure good quality, and the usual way to do so is to reverse the measurement process and measure the lack of defects.

This needs to be done in such a way that there are metrics for every stage of the SDLC, based on the four principle areas that we have established for quality control—Specification, Design, Implementation, and Integration. The reader will note that there is a phase, Definition, before Specification, which could be implicated in any defects found in the four principle stages.

However, the Definition phase is so subjective that it does not necessarily make sense to try to measure the defect rate that it might have. Such a rate might only indicate that the company has a history of attracting clients that are singularly incapable of communicating their wishes. If one is being fussy, then we could add that a high-quality team would be able to extract this information easily, and that the inability of the team to do so is a measure of that team's quality.

The problem is that the team does not possess the skills to extract the information, in which case only a team change will yield better quality, and that it is difficult to measure success based on qualitative information. The only metric that makes sense is to measure the time spent in Definition against the perceived complexity of the project, but even that will be subjective.

It is better to create a Quality Circle that reviews the result of each project with reference to the expected level of quality set out by the responsible manager. They will need to give a statement that indicates their satisfaction level, and have it reviewed during a Quality Circle meeting.

If we assume that the correct way to measure quality is by metrics that use defect removal information as a base, then we will end up with a series of metrics that give values indicating the number of defects per any one of a number of different measurements.

There is only one slight problem with using defect removal as the base measurement: it may be possible that not all the defects are found. We therefore need to try to establish what a defect is in the first place. If it is found in Unit Testing, is it a defect, or is it assumed that there will be defects when the code is initially tested?

Does a defect only become a defect if it is found during System or Acceptance Testing? Is every single item that needs to be corrected a defect, or is it only a defect when there is a palpable deviation from the Functional Specifications or Requirements Specification? Even worse—what happens if it is a defect that comes about because the Functional Specification is wrong, but is only found after the product has been handed over—should it not get more weight because it takes more resources to repair?

Answering these questions is very organization specific, and will depend on so many different factors that it becomes impossible to give a solid set of instructions. What we can do is give some examples, along with the justification for them, and let the readers implement their own sets of controls accordingly.

The first piece of advice is to use the Reporting Line as a way to establish a metric. We can therefore start our defect measurement at the first verification phase—Design Verification—and simply count the number of times that a specific team has had their Reporting Line followed back to the Design Analysis phase.

This needs to be coupled with the number of times that the Analysis has been returned via the Reporting Line to the Specification stage. This will enable us to determine whether the Design team or the Specification team made the error. Of course, they might be the same, in which case it is not necessary to try to pigeonhole a single team.

The problem might also be with the process that has led to the creation of the Specification or Design, rather than the team; all the metric can do is point to cases where there has been a defect of some kind, and measure it. Interpretation is covered as a separate topic—Monitoring Quality—later in this chapter.

Therefore, the Design Correctness Metric is the number of times that the Design Verification has failed, divided by the number of times that it has been returned to the Specification phase—one Reporting Line Hop (RLH). A value of 1.0 will indicate that each time an error has been found it has been due to an error in the Specification. More than 1.0 indicates that there are potentially more instances where the defects are a result of the Design process rather than the Specification process.

If the Design Correctness Metric (DCM) is less than 1.0, this will indicate that there are serious problems in the Specification process because it means that, per Design Verification Failure (DVF) there is more than one RLH. Since there is only one possible RLH, it indicates that the Specification team is having problems correcting the defects reported to them.

The next possible defect measurement point is at the Testing stage of the Implementation phase of the project. Here, multiple possible metrics can be taken, starting with many based on the number of lines of code. The simplest is to measure Test Case Failures (TCF) per thousand lines of code (KLOC).

We have to assume that the Test Cases are not themselves at fault, and correct them if they are. We can also create a metric, the Test Case Correction Ratio, which calculates the number of times the Test Cases need to be changed with respect to the Specifications. The TCF figure needs to be adjusted to remove the cases where the Test Case is at fault.

In addition, we also need to establish whether we want to apply the metric against the entire code base, or just against a single functional area, or object. The rule of thumb needs to be that we measure against code that we have created, in a small enough area of functionality that makes sense given the team composition.

The Test Case Failure Ratio (TCFR) is a metric that has no good or bad value, as opposed to the DCM. We can only measure a change in the TCFR to know whether we are improving. To arrive at a value that has an intrinsic meaning, we need to measure it against the RLH; the number of hops that the error has to do before the actual source is found.

So, the Test Case Failure Metric (TCFM) can be expressed as the number of TCFs divided by the number of RLHs that these have yielded. There is a maximum of two possible hops: back to the Design phase, and again to the Specification phase.

A value of 1.0 implies that, on average, the defects that are located are the result of an error in the Design process. A value that drops below 0.5 indicates that, again on average, the defects are a result of errors in the Specification phase.

On the other side, a value greater than 1.0 shows that there are more defects found than are returned to previous phases, and that the errors must therefore lie, on average, in the Implementation process itself.

The final point inside the four-phase model that we have chosen to use for quality measurements is the Testing stage of the Integration phase. We shall call each of the defects that are found as a result of Integration Testing by a slightly different name—System Failure—to differentiate the metric from the others.

A System Failure can be due to a number of different factors. It could be a result of a component failing that has been bought or acquired, the result of a component developed specifically for this project, or a result of the Assembly process itself.

Therefore, in an attempt to keep the approach reasonably simple, we will have two metrics—the SFM (System Failure Metric) and SF3M (System Failure per 3rd Party Component). The first is calculated using the familiar method of dividing it by the number of RLHs that each failure causes.

A value of 1.0 will imply that, in general, the defects found are returned to the Implementation phase. A value of 0.5 indicates that most defects are a result of the Design phase, and a value of 0.33 shows that the Specification phase caused most of the defects.

Values greater than 1.0 show that the defects are a result of the Assembly process itself, as most of the defects result in no RLHs.

It is worth mentioning at this point that many organizations will choose to skew these metrics based on the relative expense that fixing the defects incurs with respect to the number of RLHs that each requires. The thinking behind this is that it becomes difficult to tell whether there are problems when a single value is considered. The effect of multiple hops might become averaged out, and therefore we have two options:

■ Make each hop exponentially more "expensive"
■ Define metrics by hop type

Whether the former or latter is chosen will be a matter of personal choice of the organization implementing the metrics. The second will lead to the need to monitor four separate numbers in the Integration phase, three in the Implementation phase, and two in the Design and Specification phases, since one metric will need to be recorded for each hop type (return to Implementation, return to Design, or return to Specification/Definition).

Monitoring Quality

With the sets of metrics in hand, it is necessary to constantly monitor them for changes that will indicate whether there are problems that need to be addressed, and whether the general situation, by project or as a whole, is improving.

Moreover, we also need to be able to monitor the metrics by team, staff member, project, and globally. This is only possible if we have an established central point of contact with the responsibility for measuring and monitoring the metrics and their evolution. This central point is the Liaison Center.

Without involving the Liaison Center, it is impossible to accurately measure or monitor the quality of the company. For example, it may be necessary to spot trends that seem to follow a specific staff member as he changes development teams. In such a case, we can only be alerted to this if the Liaison Center has seen all the relevant documentation.

This emphasizes everything that we previously highlighted about the Liaison Center—it needs to be the only point of contact, and all handover, documentation, and result logging needs to happen through it. By doing so, while it may seem tedious at times, we can be sure that we always have a correct view of the quality and process status of the organization.

SUPPORTING THE REPORTING PROCESS

All of what we've discussed previously is impossible without clear guidelines as to how the information is to be passed back to the Liaison Center, and what they have to do with it once it gets there. With the advent of tools such as electronic mail, this has become much more streamlined and efficient.

Indeed, it is likely that the Liaison Center will use an integrated tool that will enable them to be involved in a way that does not impede progress within the project's lifespan. For example, if every piece of documentation had to be forwarded to the Liaison Center, logged, and forwarded to the intended recipient for action, as the previous discussion implies, then it would not be a very efficient way to approach software development.

Instead, some form of solution needs to be put into place that allows the Liaison Center to monitor the progress of documents through the system. We have already dealt with a large amount of theory and practical advice regarding how the key documents that are used to create the end product, as well as the product itself, are managed via the Liaison Center, but there is one piece missing.

Reporting problems and changes are the kinds of activities that require a much faster response time than would be possible if every document had to go via the Liaison Center. In fact, we need to put some mechanism in place whereby we can forward the document to the intended end recipient while keeping the Liaison Center informed.

Electronic mail can provide this functionality, but it also creates several copies of the same document, which defeats the object of the exercise. Therefore, we have to implement a solution to ensure that there is a single copy of the document, and that its status can be monitored as it goes through the various stages—from open to fixed, tested, and eventually closed.

External Documentation

In support of the preceding discussion, we also need some way to introduce the information that is required into the system from the outside. The outside, in this case, can be an internal or external client. This will be important in two main areas: Change Requests and Problem Reports.

Change Requests

A change request is raised when the functional area that has been tested is identified to not satisfy the original intentions of the client. This can be a result of insufficient or incorrect Specification or Design, or that the intentions of the client have changed over time. It is also possible that it is the result of a failure in the Requirements Analysis, but a deviation of this kind should not actually make it to the stage at which the formal Change Request document is required.

Each Change Request needs to contain a section with the project number (or other identification), client raising the change, and the stage at which the change was identified in the SDLC.

It is also important that there is a way to identify whether the Change is a result of a change in the requirements of the client, or a correction of a misunderstanding, miscommunication, or erroneous design.

This can be called the identification section of the document and exists so that the Liaison Center can monitor the various quality metrics linked to them, without reading beyond the header. The next section of the document will then contain the details of the change, which is technical in nature and intended for the Design team.

Once entered into the system, the document is stored, and only the user interface to the problem management system should allow access to it. This system needs also to be able to monitor the progress of the change through development and testing and into integration and finally acceptance.

Problem Reports

Where a Change Request can be raised when the software performs correctly, but does not satisfy the client's intentions, a Problem Report indicates that an area of functionality contains errors. There is sometimes some confusion about whether an item is an error or a change, and that decision is usually left to common sense.

The Problem Report format and processing is the same as the Change Request, except that we also need a section that specifies the severity of the problem, and the point at which it was identified in the project's SDLC, and project identification.

Motivation via Improvement

Part of TQM is to provide for continuous improvement. This can also serve as a good motivator—employees like to see their boss happy, and continuing to deliver quality will help to motivate the entire team.

However, the constant badgering of employees in an attempt to persuade them to deliver above expectation levels of quality can also lead to a de-motivating environment, so we need to be sure that they feel that they can achieve their targets, which is not going to be the case if the bar is set so high that they feel overburdened.

Case Study: Service Level Agreements

To illustrate this, let us look at a subject that should be familiar to most readers—Service Level Agreements (SLAs). For those who have not met these before, an SLA is a contract that details how a metric is to be measured, along with the expected value that the client wants the supplier to be able to deliver.

A telecommunications company, for example, may offer an SLA to their clients that states that they will be available 95 percent of a six-week sliding window period. It might then stipulate that this is calculated as being available between the hours of 2 A.M. and midnight, and that the two remaining hours lie outside the availability window. Six weeks might then be defined as 6×7 (42) days, with a total of 42×22 (924) hours.

The client might then find that one hour per day without service is too high, and they would rather have an SLA of 98.5 percent or higher. Assuming that the service provider is willing to adjust the SLA and try to reach it, with possible penalties if they do not, or bonuses if they do, the client could simply set the SLA to 98.5 percent and wait for it to be reached.

However, the chances of this happening are very slim. It could be that 3.5 percent is too much of a change, and that the service provider cannot come up with a business case to invest in the changes required to improve by this much. It is therefore a much better approach to set a modest target, and then increase it from time to time.

In this way, the service provider sees a constant improvement, which makes them happy. The client is also happy, and seeing this, the service provider will be more eager to try to do better and better. Accompanying this, they see that each modest increase in the SLA is cost effective, and have no trouble building a business case for it. Everyone is happy.

This is the approach that needs to be taken when the Liaison Center or TQM representatives find that a metric is unacceptable. Rather than setting an unrealistic target straight away, it is much better and more in tune with TQM to encourage constant improvement.

SUMMARY

This chapter was dedicated to seeing how we can feed information back from the various processes in the SDLC, which can help us to improve quality. We looked at processes, techniques, and documentation to allow us to do this, the end goal being to use the feedback to improve the product and the quality.

Improving the product is a short-term goal and means that we try to make sure we fix all the errors, implement any changes that the client imposes on us, and ensure that the entire project remains on time and within budget.

The improvement of process and product quality is a long-term goal that can only be met if the whole organization pulls together, and will need to rely on having metrics available that can tell us if the quality is improving and the current quality state of play.

It is also good to know who is responsible for particularly good or bad areas of quality—so that they can be offered bonuses or carrots. It is not a good policy to try to use a stick to persuade employees to deliver better quality results, as they may react in the opposite manner.

The better approach is to try to instill a quality culture from the start, using facts gleaned from constant quality management to illustrate areas where quality levels are good, and where they are bad. This should be backed up by rewards, which are one of the best kinds of feedback—either anticipated or given after the fact in recognition of high-quality levels that have been achieved.

19 Client Satisfaction

In This Chapter

- Introduction
- Testing for Client Satisfaction
- Planning for Failure
- Managing Client Dissatisfaction
- Summary

INTRODUCTION

The goal of quality control is to ensure that the client is satisfied with both the service that the company has delivered and the final product. Both components are equally important, since a poorly delivered product can result in the client not using some of the functionality simply because they have no idea how it is to be used, and forming a negative opinion of the developer; they may assume that the functionality they asked for is not there.

However, even if the final product contains failures and discrepancies, the client may still be satisfied with the end result if it solves the business problem that they set out when defining the requirements for the system. We could even go so far as to say that a quality product does not necessarily need to be perfect in all respects, as long as what is delivered results in customer satisfaction.

Of course, this will not always be the case; there will be instances where the customer is not satisfied with the service or product, and steps will need to be taken to address the issues that result from the negative experiences that the client has with the developer.

Before we can actually take any steps, we need to know whether the client is satisfied, which is not always as easy as simply asking them. The questions need to be couched in language that is carefully worded to avoid any misunderstandings or ambiguity, so that the client does not feel the need to be unnecessarily harsh or careful in what they say.

On the one hand, if they are encouraged to vent their displeasure, there is the risk that the meeting in which this takes place results in such a conflict that the project is abandoned. On the other hand, if the clients are reluctant to show their true feelings with regard to the developer, they may end up accepting faults in the product that lower the quality and will simply not ask the developer for services in the future.

We have split the discussion into sections dealing with how to tell if the client has had a satisfactory experience, how to ensure that, should the worst happen, there are procedures in place to ensure that these do not have an adverse impact on overall quality, and how to deal with clients who feel they have received a less than satisfactory service.

TESTING FOR CLIENT SATISFACTION

In principle, testing for client satisfaction is an ongoing underlying process that should not begin and end with the delivery of the project. The most common error made in software engineering project management is to use a project debriefing session both internally and with the client as a way to measure client satisfaction.

This is only half of the solution; it is also a good idea to consider client satisfaction at the outset of the project, too. In this way, the client's expectations can be gauged, potentially even before any contracts are rewarded.

The goal in applying the principle of continuous client satisfaction assessment is to be able to react in a timely and accurate manner, using facts as the main basis. This continuous cycle of improvement is part of the goals of Total Quality Management (TQM), and forms at least some of the basis for ISO 9000 qualification, which concentrates on project management excellence.

Besides a pure customer-facing approach to determining whether the client has been satisfied with the experience, we also would like to test that the product has achieved a high-quality status. This requires that we are in tune with the management, analysis, and technical sides of the project, as we will see.

These two aims—collect data to control response to client satisfaction, and correctly identifying the key areas of software quality—work in harmony to ensure that the primary goal of customer satisfaction is met; this could even mean that the client satisfaction issue provides limited compensation for problems experienced during the project cycle.

Pre- and Post-Project Surveys

One of the best ways to establish client satisfaction and to show the client that quality is important to the organization is by performing a survey. It does not have to be a long-winded affair, and the shorter and to the point it is increases the chance that the client will take time with it.

Essentially, the pre-project survey is aimed at finding out what the client's concerns are, and the post-project survey measures the success of the organization in providing an adequate response to those concerns.

In this way, and with interaction between the clients, project management, and project team along the way, we can establish a continuous assessment cycle of client satisfaction that will feed into the TQM and Quality Circle approach that we discussed in previous chapters.

Pre-Project Survey

The key to a successful Pre-Project Survey is that it enables clients to put their key areas of concern forward to the organization prior to starting any form of analysis or design. In fact, it might be beneficial to see what the client's expectations or concerns are in the key areas before actually tendering a bid, to avoid any problems further down the line. Key areas include:

- Cost
- Schedule and Slippage
- Responsiveness
- Value Matrix

The first item is fairly self-explanatory; it relates entirely to the overall price of the project and perceived cost. Most clients will have only a vague idea about how much something costs to create, and are very price sensitive. They will also tend to be wary of pricing that is too aggressive. It is, therefore, important to establish what they perceive as a reasonable cost for services rendered.

Schedule and Slippage relate to the importance that the client puts on staying within the time budget, with respect to the other areas of the project. For example, they may express a preference to narrow the scope of the project to avoid overrunning the schedule, but on the assumption that this will bring the overall cost down.

The third item, Responsiveness, is designed to indicate how clients expect to be dealt with when a problem occurs from their side. Most will want their e-mails or calls to be dealt with as a matter of urgency. However, they might be aware that this is not always possible, and attach less importance to this area than others—such as remaining within the time budget.

The final item is shared with the Post-Project survey and attempts to create a way of measuring the key factors of price, schedule, quality, and budgetary concerns in an attempt to see where the client attaches the most value. This is then compared with the result of the Post-Project survey to see whether the promise has been delivered on, and provides part of the answer as to whether the client is satisfied.

Post-Project Survey

After the project has been completed, a similar process to the Pre-Project Survey needs to be gone through in an attempt to establish whether the client has been satisfied with respect to the project and the opinions that they voiced in the Pre-Project Survey. On the one hand, the Post-Project Survey is a chance for the client to air their views, and point to possible flaws in the quality process; on the other, it is a way to measure changes in their point of view between the start and end of the project.

In particular, the Value Matrix might have changed: the client might, at the end of a project, be putting a different value on, for example, price, than before. If they have increased the price quotient, this may point to defects in the quality—the perceived value of what they received was less than they had intended.

Knowing why the factors causing the Value Matrix have changed is instrumental in being able to assess both the client's satisfaction and the internal processes. To do this, it will be helpful to plot the results on a graph. If we perform the Pre-Project Survey, and find that, on a scale of one to five (with one being the least important) the client rates the following as:

Price	4	Somewhat important
Schedule	5	Very important
Quality	3	Neutral
Budget	5	Very important

This tells us that Budget and Schedule are the most important factors, with Price not being as important, and Quality being neutral (expecting satisfactory quality). Generally, speaking, at the outset, Quality will be 3, 4, or 5—no clients will expect, or be looking for, poor quality results.

Once the work has been done, we then perform the Post-Project Survey, and find that, using the same scale, but with a different meaning attached:

Price	5	Excellent
Schedule	1	Unsatisfactory
Quality	4	Good
Budget	2	Below expectations

This tells us that at least, for the work done, the price was competitive. However, the schedule was not respected, and the end cost was higher than was budgeted, but quality was above the initial expectations. To get a value index, we then multiply the achieved levels by their importance:

Price	5	× 4	=	20	(max. 20)
Schedule	1	× 5	=	5	(max. 25)
Quality	4	× 3	=	12	(max. 15)
Budget	2	× 5	=	10	(max. 25)

Adding the results together, we arrive at the figure 47, of a maximum 85, calculated by multiplying the maximum possible score by the weighing factor. The client, in this case, receives an index of 55 percent, which equates to 2.76 on the scale of one to five that we have been using. They are not quite satisfied, and it needs to be established as to why that might be.

Part of the problem is usually related to cost and budgets. There are many reasons why the project may run over budget—if it is attractively priced, but runs over the initial expected estimates in terms of time and ancillary costs, then the budget will be exceeded, and the total cost might actually be more than a competitor. This will cause angst on the part of the client, and rightly so, even if the product is of the highest quality.

The Goal of Software Engineering: Quality Products

The previous section dealt with measuring the client's satisfaction with the product or project that is delivered. It gave pointers as to how to find out what the problems were, and evaluate the reasons behind them.

However, in pursuit of quality, which is part of the goal of software engineering, we need to be able to actually try to avoid the problems happening in the first place, and control their effect as we move through the project lifecycle. This requires a three-pronged approach—Management, Analysis, and Technical—each of which has a big part to play in the production of quality work.

Management

It is the role of the management team to evaluate the performance of a project, before, during, and after the project has been concluded. Management is split into three levels of responsibility: the team manager, the account manager, and the responsible manager. These may or may not map onto individual posts within the target organization, but their areas of responsibility must be dealt with by a member of staff.

The team manager needs to be able to gauge the way in which the team works together, and be able to pass information to the account manager as to whether there might be trouble in the near future. This will allow the account manager to be prepared.

The account manager is responsible for being the customer-facing contact, which is the role fulfilled by the Liaison Center in the first instance, but which will be passed on to the account manager as the project becomes established. The role is part project management, and part customer management at an abstract level.

The account manger is also responsible for keeping a close eye on the delivered quality of all projects under his control, with a duty to alert the responsible manager, who is overseeing the Total Quality Management (TQM) implementation for those areas under his control.

Abstract Analysis

To be in a position where high-quality results can be delivered on a consistent basis, analysis of the problem domain and associated areas in which quality could be compromised needs to be performed regularly.

The members of the team who understand the related business issues are best placed to do this, and will be able to relate the project and the surrounding constraints to the quality plan in a way that will enable those monitoring the continuing quality to control it.

Control of the quality in areas where analysis has shown the potential for a quality hotspot includes easy fixes such as moving more resources onto the problem, to hard decisions such as passing a functional area back to the Analysis stage in an attempt to achieve a better understanding of it, usually as a result of failures during one of the engineering processes (functional/integration testing).

Technical Matters

Finally, it should always be a point of the management to ensure that staff are assigned who have the right skills to fulfill the tasks assigned to them. Learning on the job is not an option in the pursuit of quality excellence, although a certain amount of research around a subject that the staff member is already familiar with is acceptable.

There is also a responsibility to recognize where a staff member appears not to have been correctly assigned, and to change the makeup of the team to combat possible future problems in delivering high quality as a result of a skill mismatch.

Of course, in the event of a skill shortage, there may have to be a period in which new staff are hired, but this is usually a long process, and care needs to be taken to ensure that all the skills are available, or that they can be acquired in time to avoid problems during the development life cycle.

PLANNING FOR FAILURE

There are many reasons why a project can appear to be running into problems regarding client satisfaction, but one of the main reasons is that either the organization over-promised or under-delivered on their promises. Both of these cases are avoidable, but if either should occur, things can be done to make sure that their effect is limited.

One of the key ways in which projects can build in plans for dealing with client satisfaction failure, as well as project failure, revolves around analyzing each project as it takes place and trying to identify problems that were encountered and how they were resolved, or what the effect of not resolving them was.

This will enable the company to build up a number of recommendations aimed at avoiding the recurrence of these problems, with a view to trying to ensure that they are solved ahead of time. There are three principal areas in which it is wise to try to plan for possible client satisfaction failure:

- Poor Quality Requirements Capture
- Poor Quality Implementation
- Lack of Testing and/or Quality Control Procedures

These are the areas that are most likely to cause problems both in terms of client satisfaction and in the satisfactory completion of the project. They are also the most difficult to get right, for a variety of different reasons.

Poor Quality Requirements Capture

This is a part of the cycle made difficult by the fact that the developer needs to interface with the client in an attempt to gain an understanding of the requirements of the system that will help to solve the problem they have identified. It requires that the developer try to understand the business problems, as well as deciphering the client's expression of what they say they want into what they actually need.

If the project fails to accurately capture the requirements at the outset, this will lead to problems that will reverberate through the rest of the project cycle and will likely result in a product that may be of high inherent quality but will not be an appropriate solution to the client's business or technical problems.

Avoiding Client Satisfaction Issues

Before we look at what to do if we fail to perform this part of the project correctly, it is worth taking the time to see how we can plan to avoid the problems. The first issue is when the requirements capture seems to take so long that the client becomes concerned as to the effect of such a lengthy process on the end cost of the entire project.

Simple explanation of the paradigm and processes will help to alleviate this issue—the client needs to be able to understand that time spent in the requirements capture will not so much save time in the long run as prevent future issues adding even more time to the end of the project. In earlier chapters, we discussed how finding a technical error late in the project life cycle can lead to expensive solutions, or the entire project has to be respecified, redesigned, reimplemented, and then reintegrated and tested.

The same logic can be applied to an error that stems from a misunderstanding in the requirements capture process where the client fails to convey their wishes in a way that can be understood easily by the developer. Since client satisfaction within the TQM philosophy is the number-one priority, the onus is on the developer to make sure they tease the information out of the client by rephrasing and following the logic to a variety of conclusions with which the client may or may not agree.

There may also be some frustration on the part of the client when the developer consistently fails to understand the business issues and requirements that the client is trying to explain. This can happen when the developer has moved into an area that is outside their usual knowledge domain, and can be irritating for the client when it becomes obvious that the developer is consistently misunderstanding them.

To resolve this, the client needs to be made aware of the developer's business area and expertise, and the importance of diagramming and process flow. Those involved in developing software will usually react better to a well-constructed diagram than an explanation of the business flow that is being solved.

It is also important to try to preempt this kind of problem by copious research into the business concerns shared by organizations in the client's area of business. This research will need to be presented to the team in charge of the requirements capture process in order that they understand, when going into the first meeting, what problems traditionally face businesses in the same position as the client.

Resolving Client Satisfaction Issues

It is almost inevitable that at least part of the requirements capture will fail, so we need a plan for when this occurs so we can limit the effect of this process failure. Communication will be the key to ensuring that client satisfaction remains high while there may be an area of the project that is suffering mild or severe process failures that need to be called to their attention.

Upon hearing that there is a problem, or when they notice it themselves, the client is likely to become less satisfied. They need to be aware that they should not keep this dissatisfaction to themselves, nor should they be encouraged to voice their dissatisfaction in an uncontrolled manner.

Instead, they should be encountered on their own territory, and given the chance to explain the nature of their dissatisfaction in a neutral and controlled manner. This means that a representative who is not connected with the project needs to visit the client in order to establish the nature of the concern.

It helps both sides that the representative is neutral since the client will react in a less emotional manner, and the representative will not feel that he or his team is to blame. Nonetheless, since we have established a Total Quality approach, it is a matter of pride to each individual that quality issues are tackled in a satisfactory manner, and thus that person will not be disinterested, merely emotionally neutral.

The Liaison Center will likely be the best place to find such an individual, but this requires that the organization has implemented a sufficiently wide resource base to service this part of the company. In a situation where a small- to medium-sized enterprise is implementing these guidelines, it may not be appropriate, and a neutral third party may not be available to negotiate the resolution.

In such cases, the first contact should be impersonal—via a feedback questionnaire and questionnaire reply. This shows that the developer has taken the client's problem seriously, and that they are willing to make an attempt to resolve it to everyone's satisfaction, but attempts to defuse any emotional outbursts by using a nonpersonal method of communication.

The reason why we put so much emphasis on trying to find a neutral solution is that, on the one hand, we do not have time to allow a cooling-off period of several days (or weeks), so we need to act fast. On the other hand, a face-to-face meeting will likely cause tension, which will impede the flow of information. The only way to resolve an issue where the client is not satisfied with the requirements capture process is in the exchange of information between the two parties, and any emotional involvement will impede this process.

Poor Quality Implementation

Once the requirements have been agreed, the development constraints defined, and the design built, the next critical area is the implementation of the software, and possibly hardware, solution. In fact, the implementation can be seen as a mixture of the design and programming components, and it is important that the requirements be converted accurately into a design that is then well implemented for the quality of the solution to be satisfactory.

We covered instances where we deal with errors detected by the client and possibly the developer in previous chapters, but in terms of client satisfaction, there may be cases where the simple solution of returning the erroneous piece back to the Design or even Specification stage is not a good solution.

Possible Quality Issues

Typically, as long as the design and development match the specification, and as long as the specification is an accurate representation of the requirements, which have been correctly captured, the only quality issues will be based on the user interface or interaction with external systems and processes.

Frequently, for example, the developer will believe to have found a more efficient way of doing a subsidiary task, which is brought about by a seemingly inconsequential change in the system behavior—such as altering the format of a report—without really thinking about whether it will prove more efficient once the client has had to modify (again) their way of working.

The system itself will often prove to be enough of a change for the client to deal with, without the developer trying to make the whole process as efficient as possible. Part of the problem is that solving issues of efficiency is part of the character of good software engineers, and it is almost impossible to turn it off.

Another classic example is where the organization of the user interface follows the logical progression of events in the client's processing structure but seems to group items illogically from the point of view of the system design. This may cause items to be moved around, causing the client to become disoriented when trying to evaluate whether the product meets their expectations.

Dealing with Quality Defects

Many issues based on the implementation, assuming that they are not errors resulting from poor coding or an integral misunderstanding as to the nature of the problem and solution, can be solved by simply applying rigorous processes. This will, at least, lead to the client getting what they want, even if it does not solve their business problems.

This means that there will be cases where the client gets exactly what they asked for, but not really what they needed. Dealing with these kinds of quality defects is a very difficult process for most organizations, and can even lead to cases where the project is never completed, as the client moves back to their existing system, since it has become clear that their problems cannot be solved by the implementation of a piece of computer software.

Should this be the case, the first thing the developer has to be sure of is that they get a second chance, and the way to do that is to persuade the client to reduce the scope of the project, and hence the price, to a point at which the change becomes easier for everyone to cope with. It can be a choice between this and losing the project altogether.

A phased implementation, with clear delivery points, and a plan that gives the client the option to bow out when they feel that they can extend the functionality of the new systems is also a possible compromise, and at least if some of the project can be completed to their satisfaction, then quality can be said to have been achieved.

Lack of Testing and Quality Control Procedures

Clearly, it will be impossible to deliver a high-quality product if testing and quality control procedures are not followed, or do not exist. However, as in the requirements capture phase, there can be occasions where the client becomes anxious that perhaps things are taking a little too long to conclude, and that the developer might somehow be trying to gain additional monetary reward by prolonging the end of the project with unnecessary testing.

Addressing Concerns

Again, communication and explanation will be the key to addressing any concerns that the client might have in these areas. Simulations showing why the testing is being performed in such a manner can help the client to understand the knife edge that is constantly being walked between functionality and defects.

Most clients will be unaware of the inherent danger that is brought about by the intangible nature of software, and will assume that, like building a car, if one vital part does not work, the whole engine will cease to function properly. Software, however, can work perfectly some of the time, and imperfectly some of the time, with no apparent reason, just because a single piece has not been tested under all possible conditions.

However, the client might also find that, in the later stages of the project, not enough testing has been performed, since it will appear that less is taking place due to the nature of the development paradigm being used. The reader will remember that, in the final stages of the project, the amount of testing is much reduced both in scope and in depth.

This is a result of good testing having taken place when the individual units of code were put together, and the theory needs to be explained to the client, with evidence to show that the defect rate, as measured, is comparable to other paradigms. This last is a difficult claim to prove, unless the organization, or client, has prior experience.

One final point to note is that everything will be made much easier if the organization adopts an open policy toward quality control and quality control procedures—sharing the result of quality control, showing that the Quality Circle meetings have generated real actions to address issues is all part of the client management brief.

It will also help if the interface to the quality control processes is the Liaison Center, or a third party, and not the development team responsible for creating the solution. In this way, the client will be sure that the control has been performed in a manner that is as neutral as possible.

MANAGING CLIENT DISSATISFACTION

The reason why we want to manage dissatisfaction is to gain repeat business and maintain the organization's reputation. One project that does not go according to plan will not necessarily mean the end of the relationship between the client and the developer, if that relationship is correctly managed, and any quality issues dealt with in a way that leaves the client satisfied that the developer has taken their concerns on board.

Dealing with the issues that we have discussed can often be followed back to two areas where a long-term solution can be used to address these issues:

- Better understanding of the problem domain
- Improving specification and specification review procedures

A long-term solution is necessary to avoid the same problems coming up in future projects, and these two areas are the only ones that cannot be solved by other means—design and programming problems can be dealt with by engaging staff or deploying better environments, and testing can be improved by application of a strict process and rigorous and robust methodologies.

In fact, any issues that are essentially technical in nature can be dealt with using technical solutions, but people- or information-based problems require a different approach, and it is this approach that will help to make the difference between a company with high customer satisfaction and one with fair to low customer satisfaction.

The Problem Domain

If the developer seeks to have a long-term relationship with the client, and they have not properly understood the problem domain, they will find it difficult to capture requirements or suggest solutions in that problem domain.

However, this issue is easily solved. Rather than relying on experience built up over a period of time to lead to a gradual increase in knowledge of the problem domain, all the developer has to do is recognize that there are things that they do not know, and find out how to fill the gaps in their knowledge.

Often, though, organizations are unwilling to expend resources, including financial outlay, to acquire knowledge that they feel they ought already to have, and should be easy to acquire if it turns out that they do not.

Research

It is worth investing in some research of the problem domain, including:

- History
- Existing solutions (if any)
- New developments
- Standards
- Clients/Competitors

This research is necessary because there may be things that the client leaves out in their own description of the problem domain, because they assume it to be common knowledge. In the same way, the client will probably not understand fully the organization's common knowledge in technical areas, which leads to issues when the developer realizes that something they forgot to highlight was not a part of the client's knowledge base. The developers will also have gaps in their knowledge base.

Good research, performed by the Liaison Center staff, or possibly the Librarian, or even a third party will help to build up the essential knowledge that will help in cementing a long-term relationship. It will also help in suggesting solutions that might not have occurred to other developers in the same area, or raise questions that the client had not thought to answer.

It will also help to build up a common culture between the two parties, as they will both be aware of the problems facing those involved in the client's business, and will both follow the movements within that business—project team members will probably start to show an interest in news of such movements, since they have enough background to appreciate them.

This kind of "culture matching" will also serve to build up a rapport that will extend into the inevitable business meals that will surround the project, which will help in the communication in off-the-record situations that might be of use in the project environment. It therefore serves a useful social purpose as well as a purely professional one, which creates a relationship that is far more whole than if it were based purely on the meeting room relationships that build up during the project's life cycle.

Knowledge Sharing

As mentioned previously, information needs to flow both ways, and while it is not possible to tell a client that they should research software engineering methods and practices, much can be done to make sure that the combined knowledge of all parties is made use of.

This kind of knowledge sharing will help to create a synergy that will benefit future projects, the emphasis being that it is never too late to try to understand why the client has become dissatisfied, but probably too late to repair the damage done. It is vital, then, that the knowledge sharing be performed before, during, and after the project has completed.

This sharing can quite often be completed by simply offering information about an aspect of one's own domain, and asking how that is dealt with in another. Areas that are perfect for this are quality and process control. Almost every industry has issues regarding these two points, and if a client expresses dissatisfaction with the way in which something has been done, it is quite often a good idea to try to see if they have a better solution.

This will not extend to technical issues, but might be applicable to:

- Process flows (or data flows)
- Processing logic
- Testing and Quality control

One thing that knowledge sharing is not is an invitation for open criticism, this being entirely counterproductive. It is an invitation to try to ensure that both parties can come to an agreement that will benefit them both.

Poor Quality Specifications

One area in which clients are justified in expressing dissatisfaction that should be dealt with immediately is in the specifications of the system—either the requirement specifications or the functional specifications—since these are the documents upon which the majority of the agreement for the creation of the solution is based.

On the one hand, the developer might have missed something that is not spotted by the client because there is insufficient information for them to make a judgment, and on the other, there might not be enough information to prove to the client that the developer knows what they are doing, which will inevitably lead to tension and uncertainty between the two parties.

Specification Content

Some of the classic complaints revolve around the fact that the specifications are very wordy and not easy to understand. To try to assuage the client, it is vital that diagrams be used whenever possible, and words used when a diagram is impossible. This may make for a long document, but it will be easier to understand, and help the client to be satisfied that the right system is being made, not just that the system is being made right.

Even where there are plenty of diagrams, clients do not like the overuse of technical words, and removing them upon request is one way of dealing with the issue. Adding a glossary of terms is also a good idea, as this will help in knowledge sharing, making the client feel more involved with project and more confident about the way it is going.

Finally, the longer a document is, the more a client will become dissatisfied at having to re-read it to verify that everything is as it should be. Therefore, including summaries, process flows, and diagrams representing the proposed solution is also a very good idea; again, it is all related to the notion of information sharing.

Review Procedures

Part of the reason why we try to make the documents as accessible as possible to the client is that low client involvement leads to client dissatisfaction. Helping the client to become more involved is the first step in restoring their confidence and changing the dissatisfaction into a reaction that is, at the very least, neutral. A higher client involvement during the review process will achieve this.

This means that the client must be allowed to sign off on as much as possible—which will require that they review, if not all the documents, then the procedures that have been put into place to allow the review to take place. This aspect forms part of the openness regarding quality control, and is aimed at making sure the client is happy with the way in which their money is being spent.

Finally, the lack of a third-party review process will also cause a client to be wary, and hence dissatisfied with the review process itself. Putting a third-party review process into place is not easy, however, and requires resources. The client may offer to be that reviewer, and they should be involved, but they will quickly find that they are not qualified to comment on the actual content of what is being reviewed.

Client Education

While we have dealt with the concept of information and knowledge sharing, there may be times when the client will not even possess basic diagramming skills. Therefore, there may be a need to insist that they follow some kind of education to ensure that they are equipped to be as involved as they should be in the project.

They will, of course, spend quite a lot of time learning on the job, as it were, and building up knowledge simply by the necessity of actually having to deal with the various aspects of the software engineering process on a regular basis.

The emphasis should be on future projects, and the fact that the client is building up knowledge that they will use again and again. The value-added concept can help to increase client satisfaction to the point that they will feel they have received real value for their money.

The investment by the organization in all of these initiatives is really not that great, either, especially when it is realized that, in line with the reuse policy and paradigms that we have discussed throughout the book, any training, documentation, or changes that are made are all reusable—nothing is wasted—and it all contributes to raising the delivered quality and thus increasing client satisfaction.

SUMMARY

As we have seen in this chapter, Total Quality Management is rooted in the notion that client satisfaction is the end goal in quality management. Once this is understood, it becomes clear that the way to achieve quality excellence is in recognition that clients exist in every relation between all parties involved in the Software Engineering Life Cycle.

Bearing this in mind, all of what we've discussed here can also be applied to relationships between the departments and teams responsible for delivering solutions to each other, and between the client and the developer. The application of the same policies and procedures internally and externally is part of the key to achieving quality excellence.

Issues need to be solved by a combination of communication and education, and this knowledge needs to flow both ways. This is as true for internal relationships as external ones. The members of staff who have communication issues, and it is likely that there will be some, need to be trained to be better communicators, since without effective communication it becomes impossible to solve the issues and achieve client satisfaction.

The application of these principles will result in a constant improvement of the processes and procedures that govern the software engineering tasks, which is part of the goal of managing quality under the various ISO certifications that can be obtained for organizations that have achieved process control and quality excellence.

Finally, involving the client in the TQM and Quality Circle processes will help to build confidence and satisfaction, and their feedback might also lead to useful insights into the way in which the organization has implemented their quality controls. This will be especially relevant when a client has attained ISO quality certification.

Implementation Strategies and Guidelines

INTRODUCTION

This appendix attempts to pull together the threads of discussion that have been created through this book to try to present solutions for implementing the principles that we have discussed in different corporate environments. It would be impossible to actually give details of how the various mechanisms work in every situation, so we have concentrated on three explicit areas:

- Startup companies
- Small to medium companies
- Large organizations

Some aspects of the book remain constant across all three types of company, such as those that relate to documentation standards and the various nonmovable elements of software engineering. These elements are also the pieces that are well understood, and covered in this book from a unique angle, but which have been used by organizations involved in software engineering for some time.

There are, however, some novel areas:

- The Liaison Center concept
- Emphasis on process and standards
- Code reuse and storage to facilitate minimum programming
- Quality Assurance through a service-oriented approach

Since these aspects may not have been dealt with in similar works, some readers may have difficulty mapping them onto their own organization. Therefore, this appendix should help them understand how to convert the theory into practice to improve software development practices and quality in their own organizations.

We start with a discussion of the way in which the principles can be loosely applied given differing corporate environments, essentially giving three core strategies that are then expanded in separate discussions based around the key areas of the book:

- The Liaison Center
- Development and Testing
- Quality Assurance

By gaining control of these three pillars, and respecting the standard best practices that are already prevalent in the industry, a solid foundation for software engineering can be maintained. Like all businesses, it is a case of setting up the system, and then making sure that it is constantly improved upon to maintain a consistently high level of confidence and quality.

CORPORATE ENVIRONMENTS

The first section of this appendix is aimed at offering a comparison of three kinds of environment, and the way in which the approach will be different in the way that software engineering is set up, managed, and maintained. It is a case of looking at areas that are often neglected and how this can be addressed given the resources of the organization.

We will look at the three categories that have been chosen (Startup, Small, Mature) and put software engineering into each context in turn. It is hoped that the readers will be able to begin to form a strategy for their own organization by drawing on the descriptions of these three key organizational environments.

Of course, it is never as simple as following a set of rules, even if those rules make perfect sense, since every organization will be constrained by resources and, at the root, expense. The time at which it is easiest and least costly to implement strategies and systems that are designed to enhance the software engineering process is at the start of the company's life—retrofitting is always much more costly.

However, there will be readers who want the benefit of the various techniques and concepts that have been presented in this book, but are arriving later in the game than others, and will need to try to re-engineer the software development effort to reflect the change in emphasis that this will require.

Bearing this in mind, we have adopted, as far as possible, a low-impact approach to implementing these new features, to the extent that we advocate a plan of continuous improvement, rather than an explosion of effort in an attempt to feel the benefits immediately.

The "Big Bang" approach is almost never the best way to achieve something in the IT world, since it can cause too much upset in the short term. Changing attitudes, and the way in which people relate to each other and their work, can only take place over a long period of time. In some cases, a complete change will never be possible, except by a change of personnel.

Startup

We define a startup company as one that has not yet completed a formal project for an external client, or a department that is being formed within an existing organization. The two will find themselves in a similar position, except that the latter may have more resources available in the immediate future to address the various issues that will arise as the various processes, procedures, and roles are put into place.

We assume that the team that is being put together, or is already in place, has completed a single, possible, informal project, and that they are already working together in defined roles. Should this not be the case, readers are invited to imagine the team that they had in mind, and recruit according to the new roles that that team now has to be able to fulfill.

The first, and perhaps the most obvious area in which startup software development efforts make mistakes is in under-resourcing the acquisition of tools that support the development process. This includes tools for maintaining source code and change control, testing and debugging, and more esoteric tools for team and process management.

The typical reason for this is that the staff involved are more concerned with the development environment, and competition for financial resources often puts these ahead of the concerns of the design, testing, and code management roles, which can be placed in the same basket with a lesser importance.

It is much easier, and cheaper, to put a change management system in place from the outset, even if the price seems large for the amount of code that will be placed under it; always remembering that the price includes purchase cost, training, and hardware. Most systems will function best in isolation, running on their own server: division of equipment is usually the best approach to avoid dependencies on a single device.

This is also one of the classic mistakes made by smaller companies: lacking large resources, they often try to get by with a minimum of hardware—and it is worth pointing out that machines used for development will need to be fairly resilient. Compiling, running, and debugging take their toll on hard drives, sometimes causing them to fail at inopportune moments.

Therefore, having a good backup system is also key, and again, startup companies or departments just getting under way can also leave this out of the infrastructure, usually relying on the existing system or general IT solution to provide the facility. This is acceptable, but only if the existing solution follows the guidelines for resilience and dependability that we looked at in Part II of this book.

One of the key questions that the reader will have, when looking to implement the principles from this book in a startup environment, is likely to revolve around staffing. On the face of it, it would seem that setting up a Liaison Center, an Object Library, Source Code management systems, a research team, Quality Circles, and embracing the Total Quality Management principle will require many staff that cannot be justified by a company that has, potentially, less than 10 staff members.

The solution is to use role sharing, and we cover this later in greater detail in this appendix when we turn our attention to each of the three pillars in turn—the Liaison Center, Development and Testing, and Quality Assurance.

It is important to note that when we are applying the Liaison Center philosophy in a very small organization, it is used as a framework, and not as an actual staffed office. This differs from small and medium organizations, where it becomes an actual physical office in its own right, and then in larger organizations where it becomes a department with interdepartmental links in the same manner as any other department.

There will also be an impact in the way in which quality is managed, not to mention how the various coordinating roles are split among the staff members. It is quite likely that project management becomes a function of team members, and that contract negotiation is handled by the most senior member of the personnel.

Small

The next kind of organization that we will be looking at is a small or medium-sized company or department, probably on its second or third project, and moving toward a philosophy in which development, previously performed in an ad-hoc manner, needs to be formalized in order to make efficient use of limited resources.

It can also be the case that the company is performing limited restructuring due to problems that have occurred with clients in an attempt to continue operations without undue pressure on the business situation. In such cases, it is not wise to look upon the adoption of the principles outlined in this book as being capable of saving the situation, but they will enable the company to at least make great strides toward a situation in which the company can assure clients of better service in the future.

Since the development team is assumed to have been in place for some time, having completed at least one or more projects of a reasonable size and for external clients (where external is another organization or a department within a larger organization), there will be conflicts when change becomes necessary.

In many cases, staff members will have to take on additional responsibilities, and whether these are accepted will depend on how they are presented. For some, it will be sufficient to use the change in role as an excuse to push them up another rung on the promotion ladder; for others, it may seem as if they are being burdened with extra tasks that they might not have time to service.

In such cases, the Liaison Center begins to take on a more important role, and as such needs to be accommodated appropriately. This will require some full-time staff being assigned to new responsibilities, as the Liaison Center will become a staffed communication point designed to facilitate the implementation of the various aspects recommended in this book.

Moreover, it will assume responsibility, at least in the beginning, for setting up appropriate organization and infrastructure frameworks to facilitate code sharing, and management, problem resolution, and much of the client-facing administrative work, as well as maintaining the quality plan.

These changes and new services will take time to implement, and the Liaison Center needs to have a checklist of areas that need attention, in a prioritized fashion. Therefore, if the company has a problem with an external client, such as a failing implementation project, and there is a reorganization internally in an attempt to resolve this problem, the emphasis will be on damage limitation and setting up communication channels and code management infrastructure.

If the company is restructuring prior to commencing a contract, the emphasis will be different—possibly on setting up new systems with a mind to establishing a code repository, for example. The different perspective is given by the fact that in the first case, the company was trying to get out of a tight position, and in the second, it is in preparation to get into a long-term relationship with a new (or existing) client.

We have to assume that there will be less emphasis on resource limitations than in a startup situation, although the reverse could be true; a startup might secure financing not available to a failing small company. Nonetheless, the issues that face them will be similar in terms of those areas that might suffer from lack of investment.

However, with a company that is already established, there is the additional burden of introducing change, including changes to job descriptions and existing process flows that might not be accepted by some members of staff. It might be a good approach to perform some of the tasks with an external company to reduce internal conflicts, although this can lead to problems if an external company is directly involved in giving orders to staff members.

Mature

The third and final type of corporate environment that we have identified for analysis is the mature organization. Typically, these are large companies with a tradition in IT and possibly even software development, trying to lever that experience with a new paradigm to increase efficiency and quality.

The company may, or may not, be required to do this because they have problems that need addressing. Either way, it is expected that they are prepared to put investment forward and try to establish the best framework possible, and look at quality first and expense last. This might seem unrealistic, since almost every company needs to watch the cost of new initiatives closely, but there will be cases where investment in the future wins over any arguments relating to overstretching the budget.

Nonetheless, it will be a long-term plan that is able to establish the kinds of processes and principles that we have covered in this book. Although there is always a temptation to do everything at once, while the money is there, this will usually result in an incomplete, or, in the worst case, an invalid implementation that can do more harm than good.

The issue is that a company might authorize spending on a specific area, only to withdraw it when the situation improves. Usually, this then becomes a pattern of investment when problems occur, and then not investing when the problems are solved, only to find that other problems occur.

It will probably be much easier to secure a finance plan that covers a period between three and five years, depending on the size of the organization, in which to implement the key structures that this book has described in some detail:

- The Liaison Center
- The Object Repository/Component Archive
- Quality Circles
- Total Quality Management (and relevant ISO Certification)
- Code Reuse and Prototyping approach to implementation

This represents a major undertaking for an organization that has many source code artifacts, multiple projects and teams, and separate divisions dealing with contracts, support, and customer care. Consequently, the Liaison Center becomes another department or division that interfaces in an opaque way with the other departments in the same organization.

In fact, the Liaison Center becomes a way to coordinate all the phases necessary to install the correct procedures and artifacts, including systems and processes, and is, in this case as close to the description in Chapter 1, "The Liaison Center," as one is likely to be able to get.

Getting there, however, will take a long time if there is a fractured code database, with no control over the way in which the code has been changed, integrated, and tested. In such cases, which should be rare, it will be best to start over with a new department, and merge the existing development department with it, using the Liaison Center as a coordination point.

This approach will naturally lead to the internal/external client relationship management that is part of the key to both the success of the code reuse and quality management paradigms that play a central theme in this book. Of course, it all hinges on the Liaison Center and how well it is able to integrate with the rest of the organization.

THE LIAISON CENTER

As the first pillar of the approach to software engineering that this book describes, the Liaison Center always plays a pivotal role in establishing adequate communication and managing the administrative side of the software development process. This section of the appendix looks at how the Liaison Center is set up and operates given the different organizational environments that we introduced earlier.

One thing is obvious: different organizations, with different management structures, will implement the Liaison Center in different ways. In fact, as we will see, the Liaison Center concept starts out as a framework in a startup-sized company with team sizes below 10, becomes a role-shared but managed office in small to medium-sized organizations, to a full department in larger organizations where there may be hundreds of development, testing, and design staff involved in projects.

Startup

When establishing a startup company (or an extension of a company with between 5 and 20 employees on the design, development, and test staff), the Liaison Center acts as an all-embracing philosophy rather than an organizational unit.

Each employee is likely to have a role to play in the Liaison Center, which can be seen as defining the communication between the client, the technical staff, and the design and specifications teams and developers. It is likely that the staff are already role sharing to some extent: some of those performing design or specification duties may also be implementing the solution.

If the company is a pure startup, it will be necessary to try to estimate the staffing levels required, and adding the Liaison Center responsibilities to the job descriptions will probably lead to the expansion of the staffing levels. In an existing environment, employing additional staff may become necessary, even it if is only as a consequence of promoting existing staff into positions that take on some level of Liaison Center responsibility.

The impact of this can be seen in Figure A.1, where the Liaison Center covers the topmost part of the hierarchy. It is assumed that there are three principle roles within the organization: specification (and by inference, requirements analysis), development (design and implementation), and testing.

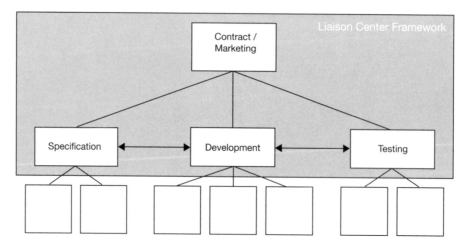

FIGURE A.1   Hypothetical startup organizational diagram.

Each of these roles has a person tasked with overseeing the projects, who communicate with each other as if each was the client of the other. On matters where the external client is implicated, this communication happens through the person responsible for contract and marketing operations. This may be the owner of the company, or the head of the fledgling development department within a larger organization.

The division of labor for the areas in which the Liaison Center is active should be done in a way that respects the capacities of individuals. For example, it is probably not a wise idea to make a member of the secretarial staff responsible for maintaining the Object and Component Archive.

The chief responsibility of the nominal "Head" of the organization diagram in Figure A.1 is therefore in maintaining contact with the client, as well as the work revolving around the contracts and limited marketing that ensures that the customer-facing aspect of the company is maintained.

Then, the Specification team becomes responsible, under the Liaison Center framework and in addition to their usual tasks, for the standards and guidelines for documentation that the company has to follow. This will also include the maintenance and operation of any tools procured for this purpose.

By the same token, the Development team then becomes responsible for Code Reuse, under the guidelines agreed with the Specification team. They are also responsible for ensuring that the Code Library (Objects and Components) is kept up to date, and that all coding guidelines are followed.

Finally, the Testing department takes additional responsibility for quality assurance, and the maintenance of the quality plan. Each member of staff should ideally also participate in Quality Circles, set up and maintained by the Testing department who then report back to the nominal "Head" for corrective action where necessary.

Small

Within a development department that has already been established, or a small software house delivering its third project, the role of the Liaison Center as an office changes slightly from that which we examined previously. In fact, it begins to become more concrete, but still more of a concept to follow than a department of its own.

In Figure A.2, we show a depiction of a minimal Liaison Center office, with four key areas of responsibility, some of which may be shared roles. For example, the Liaison Manager should be a role that is distinct from any other responsibility since he will provide a communication point between the rest of the organization.

By a similar token, the Standards Officer should probably be a nonshared role, although this is very much up to the person implementing the scheme. The reason for keeping it separate is that there will likely be an administrative or secretarial component that might be seen as too junior for a technical team member (such as a technical writer or system architect).

Unlike Figure A.1, Figure A.2 is not a hierarchical diagram; that is, it is not to be taken that the Liaison Manager is in some way senior (although he may be) to a manager of one of the Development teams. There may also be a nominal Head of Development, not shown in this diagram, who may also have one of the Liaison Center roles under his responsibility.

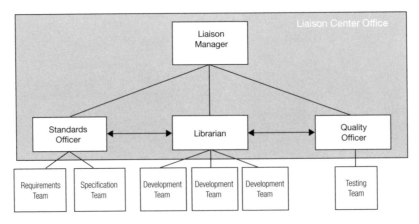

FIGURE A.2   Hypothetical minimal Liaison Center office.

The diagram shows instead a flow of information and shared or collective responsibility, where everyone is working toward a common goal. The fact that the hierarchy might look completely different in a traditional organizational chart does not mean that the Liaison Center has to be restricted by such an approach.

Additional staff members, such as developers, testers, and technical writers who make up the various teams are not shown in the diagram for the sake of brevity. They may also share some roles with the four key Liaison Center areas that are shown in the gray box.

The Standards Officer deals with every aspect of the documentation archive, as well as establishing, with the help of the Librarian and Quality Officer, standards for writing, coding, and reporting, and technical issues such as backup, restore, and change management standards.

The Librarian has the task of making sure that the Object Repository and Component Archive are kept up to date and that code reuse is held at a maximum, as well as making sure that the change management, backup, and restore processes and procedures are all operating to the guidelines set out by the Standards Officer.

The Quality Officer is responsible for making sure that the quality plan is maintained, and that participation in quality programs such as Quality Circles is kept within tolerable limits. They will also audit and control the various metrics used by the organization to measure customer satisfaction and functional quality.

The Liaison Manager has the task to make sure that all this is being done, and that the customer-facing part of the company is held to the highest possible standards. He is also the nominal head of the Liaison Center and is responsible for ensuring that the principles under which it is set up are adhered to.

Large

By the time the organization reaches a certain size, it no longer makes sense to try to farm out collective responsibility for the general operation of the Liaison Center. In fact, once the organization has more than three project teams with different clients, it is time to start thinking about setting up a proper communication point for all the various parties involved.

Since this is the ideal that we already discussed in Chapter 1, there is not very much left to add, except that the department needs to cover the functional areas of Standards and Guidelines, Information Storage and Retrieval (including Code, Specifications, Contracts, etc.), and Quality—not to mention its own secretarial team to handle direct communication with clients.

The key to operating a successful Liaison Center is in being sure that there is an interface to each area outside their level, and the possibility to bridge levels of competence where necessary. Therefore, there needs to be the possibility to communicate on a technical level with the IT department and development teams, and a way to ensure that there is correct communication toward the client in a way that they will understand.

Process and communication standards are the way this can be achieved, and it is up to the Liaison Center Manager to make sure that appropriate guidelines in place and their use is monitored and controlled. This will likely fall to the Quality department of the Liaison Center, which will need to be responsible for ensuring that a high level of quality is present in everything the organization does.

Tool-Based Alternatives

Since a small organization might not have the same resources available to establish a costly Liaison Center, replete with four or five staff (the company itself may only have five employees), certain tools can be used to augment or replace certain functions. This is also possible in cases where a large organization is not carrying out software development as part of their core business, or where management is somehow skeptical with regard to the benefit of establishing a new cost center.

For example, the Liaison Center can be reduced to a virtual framework where tasks are fulfilled by individuals under the auspices of the Liaison Center without there actually being a department or office holding that name. This is akin to the approach to Quality Circles taken in many companies where the function is fulfilled by individuals as part of their existing job, rather than nominating an individual whose primary role in the organization is as a quality coordinator.

Therefore, the Librarian function could be replaced by a suitable source code management tool, and a set of guidelines that every team leader would follow to perform the Librarian function as a shared, or collective, responsibility.

Similarly, the Standards Officer could easily be replaced by the collective responsibility of account or project managers to enforce the standards and guidelines laid out by the company, store them in a suitable document management and retrieval framework, and keep them updated with reference to their personal experiences.

The Quality Officer, as we have mentioned, is one of the easiest functions to perform as a shared responsibility, but it does require that a certain quality culture be built up within the organization.

The one function that should be considered as a noncollective responsibility is the Liaison Center Manager. While this can be carried out by a member of the organization whose primary role is not necessarily as Liaison Center Manager, it cannot be split across multiple persons, as this would defeat the object of the Liaison Center.

If these practices are put into place, the Liaison Center becomes more of a guiding philosophy, or process management tool, than an actual piece of the organization. It also enables it to be effective as a cost saver, if not an actual revenue generator, rather than an additional cost center.

DEVELOPMENT AND TESTING

The second pillar that we will look at is that of development and testing, which will likely form the core of the companies' operations. In this section, we look at how the development environment is built up with respect to the principles of code reuse and change management that we have covered in Chapters 9 through 15. This will differ from situation to situation, and again we have chosen our three phases of corporate organization to look at how implementing the principles will probably take shape in the real world.

We will also be looking at the role of testing, and how it should be coordinated depending on the size and maturity of the target organization. Like the Liaison Center, testing is a function that begins as a collective responsibility, but quickly becomes a necessary part of the organizational structure as a whole.

Startup

If a company or department has not yet performed an external software development project, it is in the best possible position to set up the various pieces of infrastructure that make up the Development and Code Library environments. This is because there is no retrofitting required—since there is no, or very little, code that needs to be integrated with the new system.

The core components of that system will be:

- Change Management tools
- Document Storage, Indexing and Retrieval
- Development and Test tools
- Backup and Restore tools

These are all vital to the correct operation of the Code Reuse paradigm that we have used as a basis for the discussion of software development in this book. They need to support the concept of pure object orientation, and need to be integrated with each other to a certain extent.

However, good process documentation and macro-style automation can help to achieve this where resources are limited, and the company has to rely on separate, Open Source solutions to perform the tasks.

In fact, it is perfectly possible to set up the compiler and development environment, a good change management system, and rudimentary documentation management system without spending any money on purchasing software applications, as long as there are no specialist or esoteric requirements.

The only resource that will be in short supply is time, which will cost money, but the savings are great in the long run. This is obvious from the discussion of object reuse and component archival in Chapters 9, 10, and 11. It is, however, quite impossible to reap the benefits of a good Object-Oriented paradigm if it is not backed up by solid systems for information storage and retrieval.

Over time, the company will be inserting source code into the system, and using it as a base for future developments, which is the key to making this process work. It is also a very good start to try to find as many standard routines and code snippets as possible when looking at a new implementation, and taking the time to ensure that they are present for object reuse at a later date.

Put another way, when starting out, it is better to spend the time building up a well-researched and stocked Object Repository. Often, it is also far cheaper to buy in and glue together pieces of code that do what is required rather than trying to develop the entire application from scratch.

Taking this approach to software development from the very beginning will mean that the repository grows over time, is complete and well documented, and that the end result should be code and applications delivered that are of a far higher quality than if each application was built from scratch.

On top of which, each component that goes into the repository will have been adequately tested within the guidelines agreed between members of the Liaison Center. It is, however, the responsibility of developers to make sure that unit testing has been done and that they do not introduce, into the repository, an item that has not been adequately tested.

The Testing department shown in Figure A.1 embodies the functionality required to perform the final validation of the system before it is shipped. This is the point at which the collective responsibility becomes the responsibility of a single department, but since the roles are probably shared with the developers, the aura of collective responsibility will continue.

Small

Small companies that are already established and are trying to use the reuse paradigm presented in this book as a way to get more out of their development resources while also planning for the future will find that it is an expensive but ultimately profitable task.

In fact, it is likely that simply picking up the existing code base and trying to shoehorn it into a well-defined system of processes and application software will prove counterproductive.

A better approach, but one that will meet resistance due to being resource intensive, is to reverse engineer existing code, and repackage it as neat little objects, which are then inserted into the repository as if they had been acquired externally. This will cause some level of reengineering, and will probably lead to almost all the code that has been written to date to be respecified, redesigned, and reimplemented to make sure that the right foundation is made for future object reuse.

The effort will be well worth it, but unless the company has prepared for this kind of approach in advance, it will probably also be too expensive. There is a compromise, which involves putting the systems and procedures into place, but only using them for future projects.

The catch is that, in order to benefit, the existing code has to be integrated at the same time. Therefore, the process becomes one of:

1. Deciding the nature of the component required.
2. Looking for it in the repository, then in the code base, then elsewhere.
3. Using, updating, and placing it (back) into the repository.

In this way, we satisfy the requirements of being able to populate the repository with existing code, while also writing as little code as possible. The key is in step 2 where we look in the code for an object or snippet that contains the required functionality, and remove it if it exists, with a view to placing it into the repository as a new object.

However, should we not find an appropriate piece of code, we then need to look at large for existing modules that perform the task we have in mind. Only as a last resort do we write the code ourselves. The time spent in research is assumed to balance the time that would have been spent in development and testing.

The issue arises when we need to ensure that the existing application that we are poaching code from remains operable. In such cases, we need to perform an additional step to reintegrate the new functionality with the application such that it functions in the same way as before.

As we mentioned at the start of this section, it is a long process that is difficult to put into place once the software development environment has been in place for some time; of course, if the application is a throwaway, then the issue of retaining a functional application after the fact is less important, and we can archive it as is in case it is needed at some point in the future.

This is also the point in an organization's development efforts where testing becomes more of a specialist department role. Individual developers will still be responsible for making sure their code works as advertised, but a department that takes overall responsibility for testing activities will be needed simply due to the volume of code that is expected to be connected together.

Since it is generally agreed that the complexity of the testing role will increase with a larger number of projects that may be diverse and probably sharing components, having a dedicated testing team, as shown in Figure A.2, and reporting through the Liaison Center, is probably the best way to ensure a consistently high level of quality.

Large

As the reader might have guessed, at the point where the organization can be considered large, there is no possibility to try to reverse engineer the entire code base to fit into a new code management paradigm.

If we assume that the development department has organically grown to a point at which it is necessary to try to formalize code reuse, but is well supported in terms of tools and change management, then we have only to deal with the issue of the Object Repository or Component Archive.

This is a fair assumption because it is unlikely that a company that has managed to deliver good-quality software on a consistent basis does not have the tools in place to guarantee this to their clients. Hence, we can be almost certain that the organization is merely trying to capitalize on this and operate in a way that is more efficient and even higher quality.

Therefore, the best approach will probably be similar to that for Small companies—to try and regroup functionality on an as-needed basis under the newly created library. Of course, the Librarian could also work his way through the existing code base and try to create packaged components out of the various pieces that have been written, and place them in the Component Gallery for future use.

This approach will probably work better for larger organizations because it is assumed that a certain level of reuse already exists, where development teams are in constant, informal, communication, and it is merely a case of formalizing these channels.

As far as testing is concerned, the emphasis has to move toward a strategy for testing and test management with respect to the new paradigm for object reuse. That is, the programmers will need to be aware that all the code they write will be inserted into the repository, and therefore they should make sure that it has been thoroughly tested before they do so.

QUALITY ASSURANCE

Finally, we need to say a few words about how the third pillar, which is also roughly equivalent to the third part of the book, is implemented in our three categories of organization. Quality Assurance is more than just testing—testing is merely a measurable indicator of success of a technical implementation.

Quality Assurance is a multidiscipline task that stems from the principle that the client comes first. In fact, the emphasis of this book has been on creating software engineering departments and organizations that treat the way software is created as a process that delivers a service, rather than a project.

If we look at the Open Source environment, we see that the actual application software is usually available for download. The collection of applications, from operating systems to word processors, development environments to Web servers, is vast, and covers almost every aspect of information technology.

They cannot be sold. In fact, as we saw in our discussion of the various Open Source licensing restrictions, it is often illegal to try to sell them. However, the delivery of a service that has an Open Source product at the core is perfectly acceptable.

Cases in point include SuSE and other distributors of the Linux operating system, as well as the numerous consultants who regularly use Open Source applications to help them deliver their own service.

This is a trend that this book capitalizes on, by emphasizing process and construction from existing parts over programming from the ground up. Hence, it becomes about delivering a service, not creating a product. The product is the service, and is bound by quality assurance rules that have been in operation for many years.

Startup

In terms of the startup, formal quality assurance measures are often overlooked due to the compact nature of the environment. It seems to make little sense to speak of Quality Circles in a context where the whole organization could probably be represented by a single circle.

However, the formal procedures are still valid. Staff meetings can become part Quality Circle, and the guiding philosophies of Total Quality Management remain the same—in fact, it should be far easier to monitor the quality plan in a smaller organization.

Client satisfaction still needs to be measured, even if it is only to be sure that they are happy with the work, and will consider hiring the organization again in the future.

Of course, contacts within this context will be less formal than in a large organization servicing large projects for similarly sized external companies. However, it is worth putting in the extra effort to obtain ISO Quality Certifications, as it will stand the company in good stead for bidding on future contracts.

Small

As a small company, it becomes more vital to try to gauge how it is performing in terms of service to the client and internal quality control than to attempt to gain official certification. It is assumed that there has been at least one satisfied customer, and if the reengineering of the processes involved is a result of dissatisfaction, it is probably not the best time to start a process that will not address this issue.

In the spirit of client-first quality assurance, it is this that must be addressed, and to do so will require that the clients are invited to air their views in a frank manner. We dealt with the processes and procedures in Chapters 18 and 19, and will not reproduce them here, but it is clear that if setting up a dedicated quality program is in reaction to client dissatisfaction, then the cause needs to be found.

However, if the company is simply trying to instill internal confidence in the quality of the service they deliver, or as a general quality improvement measure, they can take time to ensure that the infrastructure is in place prior to trying to evaluate the current position and how it can be improved.

It is sensible to try to obtain outside help when attempting to redress issues of process quality in the first instance, since those working closely within the organization will not tend to be objective enough to do the task justice.

Large

A large organization needs to approach the Quality Assurance issue with reference to existing structures that are in place. Reusing experience from other divisions is vital in trying to establish a strategy that will deliver consistent results; this will save a lot of time in trying to find the best way to implement the key principles of Quality Circles and Total Quality Management.

Moreover, these will not work in their entirety unless the entire organization is involved. This means that, more than ever, if a company tries to put these principles into operation once the organization has reached a certain size, the financial resources required will be prohibitive unless an approach is adopted that brings the measures in over a fairly long period of time.

It is suggested that the first facet to be put into place is a system of scoring customer satisfaction and noting those areas where there are deficiencies. This will allow the responsible person to put measures into place, such as Quality Circles in areas that are underperforming, while drawing on any informal measures that might be present in areas that are performing well, or at least better.

Eventually, with the right level of employee participation, it should be possible to engineer Total Quality Management without an explicit top-down directive, which is the worst way to try to instill this kind of support. If people are being forced to do something, they are much less likely to want to do so than if it is a result of pride in their work or even peer pressure.

SUMMARY

In the final analysis, it is clear that attempting to implement the various pillars and principles is no easy task if no groundwork has been done. The simplest solution is quite often just to try to start again, without any attempt to directly integrate the existing situation.

However, at some point, the employees at least are going to have to be integrated, even if their work is not, and therefore their support has to be solicited. This will mean that at least one member of the staff has to become a convert, and try to make sure that the rest of the organization follows suit.

This will be easier to manage with 10 employees than with 50 or even 100. At the end of the day, though, those who do not become part of the dream will have a negative impact on the overall success of the organization trying to adopt these measures.

There will also be cases where the mantra of trying to treat the delivery of software comprising objects connected by very simple logic under the auspices of treating the whole operation as a service to a client will not work. In such cases, although the general principles of the Liaison Center and Object Repository can be respected, it will be difficult to divorce the procedure from one in which a product is being created.

This will not impact the way in which the client is treated, but may mean that there is more scope for lower than anticipated quality levels due to an increased reliance on nonstandard parts. The trick is to try to ensure that the risk of this becoming a problem is minimized—the role of the Software Development Manager.

Appendix

B   About the CD-ROM

The CD-ROM included with *Corporate Software Project Management* includes a set of skeleton documents referred to in the book, as well as some useful software applications, and links to useful Web pages. Most of the content is available from the "index.html" file available in the root of the CD-ROM.

CD-ROM FOLDERS

Applications: Contains sample applications that have been referred to in the book, as well as some which the author believes to be of use in planning and executing software development projects.

The commercial applications "MindJet MindManager X5" and "RSM" are trial versions, limited to 21 days and 10 files, respectively.

Images: Contains all the images in the book by chapter. These are not available from the root "index.html" file.

SkeletonDocuments: Contains a set of templates, in rich text format, which can be used as skeleton documents for starting up new projects.

OVERALL SYSTEM REQUIREMENTS

- Windows NT, Windows 2000 Professional Edition, or Windows XP
- A Web browser
- Pentium IV Processor or greater
- CD-ROM drive
- Hard drive
- 128 MBs of RAM, minimum 512 recommended

Other software packages may have requirements which differ from those above, but in no case should they exceed the recommended hardware specifications.

MindJet MindManager X5 (*http://www.mindjet.com*)

To install, run the SETUP.EXE application in the Applications/MindJet folder.

Additional System Requirements

- Windows NT® 4.0 SP6/2000/XP Professional, Home or Tablet PC Edition
- MB disk space
- x600 resolution or higher, 16-bit / 65Kcolors or higher

OpenWorkBench (*openworkbench.org*)

To install, run the owb.1.rc3.exe application in the Applications/OpenWorkBench folder.

RSM (*mSquaredTechnologies.com*)

To install, extract the rsm.zip file located in the Applications/RSM folder.

Subversion (*subversion.tigris.org*)

Installation instructions are contained within the README.txt file located in the svn-win32-1.0.5.zip archive in the Applications/Subversion folder.

WinCVS (*www.wincvs.org*)

To install, execute the SETUP.EXE application contained in the WinCvs13b17-2.zip archive file.

Index